Collectible Match Holders

for Tabletops and Walls

Jean and Franklin Hunting

Schiffer Publishing Ltd®

4880 Lower Valley Road, Atglen, PA 19310 USA

Dedication

This book is dedicated with love and appreciation to our son, Franklin Randall Hunting. Randy's expertise with the computer was invaluable; he rescued us many times. His good natured patience kept us going through the trials and tribulations of modern technology.

Library of Congress Catalog Card Number: 98-86995

Photo layout by Randy Hensley
Designed by "Sue"
Typeset in Bodini BT/ZapfHumnst BT

ISBN: 0-7643-0691-X
Printed in China

1 2 3 4

Published by Schiffer Publishing Ltd.
4880 Lower Valley Road
Atglen, PA 19310
Phone: (610) 593-1777; Fax: (610) 593-2002
E-mail: Schifferbk@aol.com
Please write for a free catalog.
This book may be purchased from the publisher.
Please include $3.95 for shipping.

In Europe Schiffer books are distributed by
Bushwood Books
6 Marksbury Avenue
Kew Gardens
Surrey TW9 4JF England
Phone: 44 (0) 181 392-8585; Fax: 44 (0) 181 392-9876
E-mail: Bushwd@aol.com

Please try your bookstore first.

We are interested in hearing from authors with book ideas on related subjects.

Contents

Acknowledgments

Our heartfelt appreciation goes to the many people who furnished us with photographs, information, and, above all, encouragement. The best aspect of doing this book was getting acquainted with others who love collecting match holders as much as we do. It was most rewarding and a lot of fun. The people listed below were very generous with their time, knowledge, and patience; their contributions enhanced the book considerably.

We thank you all.

Barbara and Steve Aaronson. "The Victorian Lady." Northridge, California.

Mattie and Allen Alperton. San Francisco, California.

Barbara and Will Andrews. "Andrews Antiques," Sacramento, California.

Judy Andrychuk. Roscoe, Illinois.

Bonde' A. Bliven. "Bonde' Antiques." Carmel, California .

Barbara and Kenneth Bloomfield. Belvidere, Illinois.

Marsha and Dell Breon. Modesto, California.

Agnes Chambers. "Before My Time" Antiques. Canoga Park, California.

Johnny Crabb and Merlin Nomann. Modesto, California.

Diane and Roy Farriester. Modesto, California.

Linda Fisher and Andreas Kazos. "Fisher Gallery." Sacramento, California.

Ed and Lenore Fuetz. "Trails End Antiques." Tazewell, Tennessee.

Morris Gersten. Scottsdale, Arizona.

Louise Kossmann. "Kossmann Antiques." Fresno, California.

Marti and Bill Kurschat. Fullerton, California.

Lorraine and Myron Leveen. Holbrook, New York.

Mr. and Mrs. Dick Marsuobian. Fresno, California.

Maria Mesing. "Maria's Antiques." Hemet, California.

Glen Movey. "Treasure of The Sierra Madre." Fresno, California.

Patricia Sanders. Fort Dodge, Iowa.

Warren A. Shuman. Denver, Colorado.

Sunny Lenzner. "Sunny Lenzner Antiques." Escondido, California.

Pat Mizrahi. "Mizrahi Antiques." Sherman Oaks, California.

Eve Naslund. El Cajon, California.

Dottie and Onie Sims. Whittier, California.

Jesse Williams and Jim Ellwood. "Irontiques." Scottsdale, Arizona. A unique shop, they usually have a nice supply of match holders.

A grateful thanks to Jacqueline Hunting for critiquing the manuscript.

Also, a special note of gratitude to Richard Osterberg who proofread the manuscript. His helpful suggestions and encouragement throughout were most appreciated.

Introduction

The first match holder in our collection was purchased on a trip to New England in 1973. We spotted a match safe, "acorn on a pedestal," in a small antiques shop in New Hampshire. We were intrigued by the tilting container in the form of an acorn and the fact that it had a patent date of January 21, 1862, on the side. Of course, we purchased it, and it has been on our coffee table ever since, along with several others acquired over the years.

We have been in the antiques business since 1972, Hunting House Antiques, specializing in sterling silver. We have lived most of our lives in California, but, in 1963, Franklin, who worked for Westinghouse Electric Corporation, was transferred to Baltimore, Maryland. We loved Maryland and enjoyed the opportunity to go antiquing in the east, especially Pennsylvania and New England. However, our families were in the west, so when Franklin retired in 1982, we headed back home – first to Scottsdale, Arizona, for a few years then back to California. We have traveled coast to coast many times and have shopped for antiques along the way. We have participated in antiques shows on both coasts and places in between. We have been exposed to literally thousands of antiques of all sorts and we were constantly looking for old match holders.

Our interest was primarily in American match holders. We discovered that match holders were seldom hallmarked, so it is impossible, in most cases, to ascertain when and where they were made; however, by studying American patents and talking to other collectors we were able to fill in some of the gaps. We especially prized the ones that were marked with the name of the manufacturer and a patent date. We eventually acquired over two hundred match holders, for the wall as well as for the table. We could not stick to our desire to collect only American match holders, just couldn't resist a rare or cute holder regardless of its origin. We displayed them all over the house, overdid it a bit, had a large assortment hanging on the kitchen wall, the entry hall, and places in between. The coffee table accommodated a few and a curio cabinet housed the rest.

Over the years, we have noticed a definite decline in the number of match holders on the market. Match holders are wonderful to collect; they are small so they do not require a large area to display them, and they are intriguing in their design, so many people are attracted to them. Match holders are rapidly disappearing into private collections and museums. Along with scarcity and inflation, they have increased in value a great deal over the past few years.

Many of the pictures in this book were taken over twenty years ago. Through the years, we have appealed to our fellow antiques dealers to allow us to photograph the ones in their booths, at the shows, or their shops. Private collectors also contributed greatly: they were very generous with their knowledge and pictures. We have described the match holders, taken measurements when possible, and identified the maker and country of origin, when known. The majority pictured are American made even though they are unmarked. Photographs not attributed to others are in the authors' collection.

Matches have been a blessing to mankind, it is hard to imagine what life was like when making a fire was done with flint and steel. We have included a brief history of the invention of matches and match containers, and a section on United States patents. Where would we be without those ingenious people who designed and produced such wonderful little match holders?

Value Guide

It should be remembered that a value guide is just that, a guide. It is not intended to set prices or to be the final assessment of the value of an item. Instead, it is intended to give a broad range of retail prices that can reasonably be found within the continental United States. A match holder in superior condition will certainly be valued on the higher end, one of average condition valued somewhere in between, and one of poor condition for considerably less. Additionally, a beautiful, rare match holder will certainly be valued higher than a plain, common one. Prices in a value guide are merely rough estimates of the current price and are subject to change. We arrived at these prices using our years of experience as collectors and our twenty-five plus years in the antiques business. We have also consulted with other collectors and dealers across the United States.

It appears to us that the Midwest and the East Coast of the United States have a larger supply of match holders. Since they are more plentiful in these regions, they are usually less expensive than ones found in the West. Although match holders were made in great abundance, they are becoming harder to find. Scarcity, coupled with inflation, has caused a significant price increase.

Auction houses publish a list of merchandise to be auctioned off and give an appraisal of each piece. Seldom do the expected prices match the actual selling price and much depends on what the individual is willing to pay for a desired object. The same is true in any field where there are so many variables, such as age, rarity, beauty, geographical location, and the desire to own an object. A department store has to set its prices in alignment with its competitors or the customer will go elsewhere because the same new merchandise is widely available. With antiques it is not so simple. You cannot always find another identical item, and if you pass up a wonderful example just because the value guide has it listed lower than the asking price, you may be making a mistake. By the same token, if a nice match holder shows up at a garage sale or a flea market for a fraction of the listed price, it might be your lucky day.

Due to the many variables that affect the price, match holders may be found selling for more or less than the estimated price listed in this book. Neither the authors nor the publishers assume any responsibility for the losses that may occur from using this guide.

Chapter One

History of Matches

Making fire was a difficult task. Whether it was rubbing two pieces of wood together, as primitive people did, or striking flint against steel, it was a time consuming project. Anything flammable, wood shavings, leaves, or bits of fabric, was placed in a metal "tinder box," and, with vigorous striking of flint against steel, a spark would eventually ignite the contents. To get a good fire going, it was necessary to blow on, or fan, the smoldering tinder, which took a half hour or so to get a sustainable flame. Matches, as we know them today, are a triumph of research and development. A small wooden stick tipped with chemicals, which when struck will burn for just seconds, has brought untold benefits to mankind.

The first friction match was made in 1827 by John Walker, an English chemist, who took a wooden splint and dipped it into a mixture of chlorate of potash, gum arabic, and sugar. He then drew it through a piece of folded sandpaper to create a flame. In 1831, Dr. Charles Saurian, a French scientist, invented the first successful friction match by substituting a phosphorus compound for the chemicals that were used in England. On October 24, 1836, Alonso Dwight Phillips of Springfield, Massachusetts, registered a patent (number 68). It was the first friction match to be patented in the United States and was made entirely by hand.

These matches became known as "loco focos" or later "lucifers." The odor from the lucifer matches was so offensive that "parlor" matches were invented to keep the air free of pollution. Because parlor matches contained chlorate of potash, which is highly flammable and explodes in a shower of sparks when struck, carpet fires often resulted. In Europe, the "Vesta" match, a miniature candle with a match head, was used extensively. It was a more delicate match, and when struck the head was apt to fly off.

The lucifer match was used in the United States until the invention of the safety match, which would ignite only when struck on a special surface. The lucifer was highly combustible and ignited with the least amount of friction. Men would often strike a match on the seat of their trousers. Not only would this result in a shiny patch, but it had the potential for disaster. Another practice was to use the fingernail of the thumb to flick the head of a match into a flame. This could become a painful experience if the match head were to break off of the wooden splint and lodge under the nail complete with flame.

When the friction match came into common use, the individual small manufacturer could not meet the demand and a more mechanized method of production was needed. In 1881, twelve pioneer match companies combined to form the Diamond Match Company.

In the manufacturing of early matches, a deadly form of white phosphorus was used. When inhaled, this gas caused a horrible disease, known as "phosphorus necrosis," which killed many factory workers. It attacked a decayed tooth and then spread to other teeth and the jaw bone. Eventually, it lead to death. In 1911, at the Diamond Match Company, William A. Fairburn, a mechanical and chemical engineer, solved the problem of this disease by substituting the French patent for sesquisulphide of phosphorus, a nonpoisonous chemical, for the poisonous white phosphorus. With the addition of chlorate of potash to the mixture, a greatly improved match came into being. This new combination of chemicals allowed for the creation of the first nonpoisonous safety matches in the United States. By raising the temperature of ignition by a hundred degrees and soaking the wooden splints in a chemical solution to deaden the afterglow, the matches were made even safer. The odious smell and the reeking smoke from the old lucifers became things of the past. In a humanitarian gesture, the Diamond Match Company generously donated the patent for this life saving process to the American people.

Chapter Two

History of Match Holders

Before the advent of the safety match, which would only ignite when struck on a special surface made for that purpose, matches were extremely volatile, igniting with the slightest friction. "Loco foco" and "lucifer" matches were unpredictable in the extreme. When struck, they emitted a cloud of noxious smoke and a foul odor. To store these early matches, a special container was needed to protect them from starting accidental fires.

Although the term "match safe" refers to the containers used to keep these matches from igniting accidentally, the terms "match holder," "match container," "match box," and "match striker" all describe the same article. This has created some confusion. Match safes, in our opinion, are closed receptacles such as pocket match containers or holders with a lid that can be closed. On the other hand, "match holder" may refer to any container that holds matches.

Match holders were produced in every available material: cardboard, glass, iron, porcelain, pottery, silver, tin, wood, and bronze. White metal, an alloy of antimony, copper, and tin, was used a great deal. Because tin was less expensive than most other materials, it was used extensively for manufacturing match holders for advertisements. Since tin match holders were inexpensive to manufacture, costing just pennies, they were often given as gifts to customers. A match holder bearing a shopkeeper's name and hanging at eye level was an excellent way for a merchant to advertise his services on a daily basis.

Many match holders were made as souvenirs and sold at fairs and other places where people congregated. Yellowstone National Park, the Liberty Bell in Philadelphia, Mount Rushmore, and Niagra Falls are a few of the attractions commemorated. For the Florida souvenirs, an alligator and oranges were often featured somewhere in the design.

In accordance with the old saying "necessity is the mother of invention," American inventors produced some of the most artistic, interesting, amusing, and safe match holders, using all kinds of materials. Every imaginable form was used as a subject. There were match holders to hang on the wall, to sit on the table, and even to carry in one's pocket. Match containers were also combined with other items such as gas light fixtures, calendars, inkwells, cigar cutters, thermometers, and bells.

The subjects depicted on match holders vary; any and everything was used. There were animals of every sort. Dogs, mules, camels, donkeys, monkeys, cats, and alligators were among the most popular. People from all walks of life, the aristocrat, the bum, and the court jester, were also displayed. Objects in daily use, such as shoes, boots, upside down hats and umbrellas, hollow tree trunks, buckets, and baskets, are just a few of the items that were used to function as match holders. The "lucifer" match, which put out a foul odor and a choking cloud of smoke reminiscent of fire and brimstone, inspired designs using the devil as the main theme. Effigies of rats and mice, often found chewing on the phosphorus match heads and causing many fires, were also used to depict the safety of the match container from rodents. Whether a fancy porcelain container on the parlor mantle or a plain iron one by the cook stove, match holders were indispensable for safely storing unstable matches.

With few exceptions, match holders display a rough, ribbed, or corrugated area for striking the match. The striking surface is usually prominently displayed on the front, the sides, or bottom of the match container, or often times on all three places. In some examples, the striking surface is incorporated into the overall design, i.e., the rough bark of a tree could be used as the match striker. In many cases, the striking surface is located under the lid of the container. Many porcelain match boxes have gone unrecognized for their true purpose because the ribbed striker is in such an inconspicuous location. The lack of a striking surface indicates the item probably served another purpose. If the design is smooth with no place to strike a match, even if it appears to be a match holder, it probably is not. However, there are exceptions. There are the match holders in the form of boots that do not have a striking surface. Many companies, in the Victorian days, made silver plated match holders without strikers. These were advertised in their catalogues as "match or toothpick holders."

When shopping for match holders, the collector should be aware that reproductions abound, and it is a good idea to learn to spot them so you will not pay the price of an original. Reproductions do have their place and an advantage is that they should sell for considerably less. Reproductions were not necessarily made to deceive, but were made to provide a replica of a desired, but scarce, item. Fortunately, some companies signed their reproductions. Holders marked "Iron Art" or "Wilton" are modern day reproductions of antique holders. In their 1976 catalogue, Wilton Products, Inc. advertised hand cast reproductions of Early American holders. Unfortunately, by ignorance or mendacity, some reproductions are passed off to the unsuspecting collector as an antique. Here's a good way to tell the difference. Iron will rust if exposed to the elements. However, the difference between old rust, rust acquired over a period of years and new rust, rust purposefully created to give the impression of age, is new rust is much brighter in appearance and looks to be what it is — new. An old iron match holder is usually well cast and smooth to the touch, while a reproduction is often coarse and not as fine in appearance. Another easy way to identify an original iron match holder is to look at the back side. Sometimes the residue of Japanning, a brownish varnish used in the early days of manufacturing, is still visible. In addition, inspect the inside of the container. If it has not been repainted, signs of aging will show. The best way to familiarize yourself with the difference between a reproduction and an original is to see and handle as many old match holders as you can and talk to knowledgeable dealers.

Condition is of utmost importance. There is an old saying, "if it's chipped, it's cheap." Iron is probably the easiest material to take for granted, because we do not expect it to break. Although iron is extremely resilient, it can be broken when dropped. So, make sure to look the container over carefully to make sure no pieces have been broken off. Painted match holders lose some of their value if the paint is badly worn. A silver plated piece in need of new silver will receive a much lower price than one in mint condition. Examine glass and porcelain holders carefully. The slightest chip or crack will diminish the value of a piece. Although it is often impossible to find perfection in old pieces, a little caution should be taken to avoid a really bad purchase.

When automatic pilot lights were added to gas stoves, electric stoves replaced the old wood burning ranges, and, with the invention of the safety match, the need for a secure place to store highly combustible matches became a thing of the past. The new safety matches, with the striker on the side of the box, could be kept in the carton in which they were purchased. For the most part, match safes had outlived their usefulness. They had been invented for the unstable early matches. With the invention of the new safety match, they became useless and many were thrown away or relegated to the attic.

Today, a display of old match holders, hanging on the wall, displayed on a table, or in a curio cabinet, adds to the decor of any home and is a great center of attraction. Match holders can still be used as intended, but must contain "strike anywhere" matches; safety matches will not work. A match holder hung by the fireplace is practical as well as pleasing to the eye.

Match holders do not have to be more than one hundred years old to be worthy of collecting. In time, newer match holders will become antiques as well. The easiest place to find match holders is at antiques shows and shops. Here, the dealer has done the leg work for you. If you enjoy the hunt and have the time, flea markets, garage sales, estate sales, and auctions can be good sources too. Also, remember to check your own attic. Although match containers were made in great abundance, they are becoming scarce. The population of the United States has increased immensely, so more people are collecting in a limited field. Remember, if old match holders were easily found they would cease to be of interest to the collector.

Chapter Three

Match Holders For The Table

Beautiful bronze match holder for the table. A court jester is gracefully balanced with one foot on a ball while juggling a ball in each hand. His arched back supports the match urn. The round base has a ribbed surface for striking the match (c. the last quarter of the nineteenth century). Base 3" in diameter, 4 1/2" tall. $300-400.

Above and left: Alternate side views of the match holder.

Match holder for the table in the form of a cottage. Embossed around the edges of the base is "COMPLETE MATCH SAFE PTD JULY 13, 1883." The roof lifts off and the matches are stored inside. The striking strip is located on the roof. Iron. American. Base 3 3/4" x 4 3/8", 4 1/2" tall. *From the Alperton collection*. $200-300.

An old top hat with a hole on the side is the container for matches. The bottom of the hat is roughened to provide friction for striking the match. England (c. the early 1900s). Brass. Base 2", 1 1/2" tall. *From the Kurschat collection*. $60-85.

Probably the first known functional match safe. Formerly an eighteenth-century tobacco box. The box was converted into a match safe when tobacco packaging improved and the need for tobacco containers was eliminated. In the mid nineteenth century, the words "MATCH SAFE" were printed in gold on the front and a dog was painted on the back. Iron. American. 3 3/4" x 5", 6" tall. *From the Alperton collection.* $300-400.

The lovely figure of a young woman ice skating. She is attired in Victorian clothes, a long skirt, a jacket, and a hat with a feather. Her hands are kept warm in a muff. The match container is positioned on the side with strikers on each end. The figure is made of white metal and stands on a round alabaster base. American (c.1875). 5 1/2" in diameter, 9" tall. *From the Kurschat collection*. $450-550.

Cast iron match holder for the table in the form of a turtle. The round container on the turtle's back holds the matches. The ribbed sides are used as the striker. American (c.late nineteenth century). 2" x 3 1/4" wide, 1 1/2" tall. *From the Kurschat collection*. $125-145.

Match holder with the figure of a man who appears to be hailing someone. There are tassels on his cape, front and back, and on the top of his boots. A basket behind the figure holds the matches. The rough sides of the basket are used to strike the match. France or Belgium. White metal figure. Iron. Base 4 7/8" in diameter, 8 1/2" tall. *From the Kurschat collection*. $295-350.

Open view of the brass match box.

Brass match box for the table consisting of a pair of acorns resting on a bed of oak leaves. The lid is hinged and opens to expose two compartments. The striker is located on the edge of the leaf in the back (c.1880). 4 1/4" x 5 3/4". $100-150.

Match safe in the form of an acorn on a pedestal. The lid is attached to two supporting arms which allow the urn to tip open, exposing the matches inside. The base is decorated with oak leaves and acorns. One section of the base provides a rough area for striking a match. Engraved on the upper edge of the urn is "PAT' JAN 21, 1862." Iron with a bronze patina. American. Base 3 1/4" in diameter, 4 3/4" tall. $175-275.

Cigarette and match box with a hinged lid. The inscription on the left side reads "C CY COHEN IF IT IS INSURANCE I CAN PLACE IT" and on the right side is "6TH & WASH STS READING PA NOTARY PUBLIC." The rectangular block on the right side holds a small box of matches. Iron. 4 1/2" x 7 1/8", 2" tall. *Courtesy of Irontiques*. $175-225.

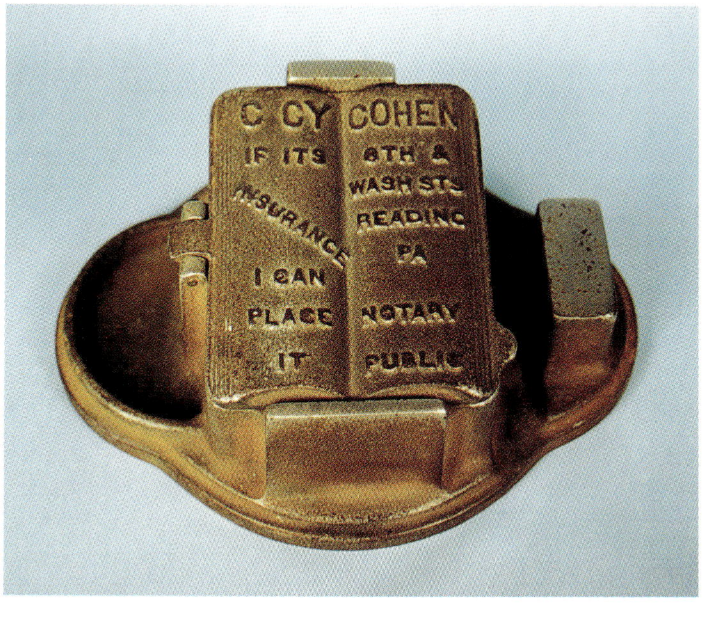

Flat match box with a hinged lid. Straps decorate the top of the box. The bottom is ribbed to function as the match striker. #1833 is inscribed underneath the lid. Iron. 2 1/2" x 4". $65-85.

This match holder for the table has a separate lid which has a round brass knob on top. There is a pierced design around the sides of the box. The striker is located under the lid. Iron. American. Patented 1875. $95-150.

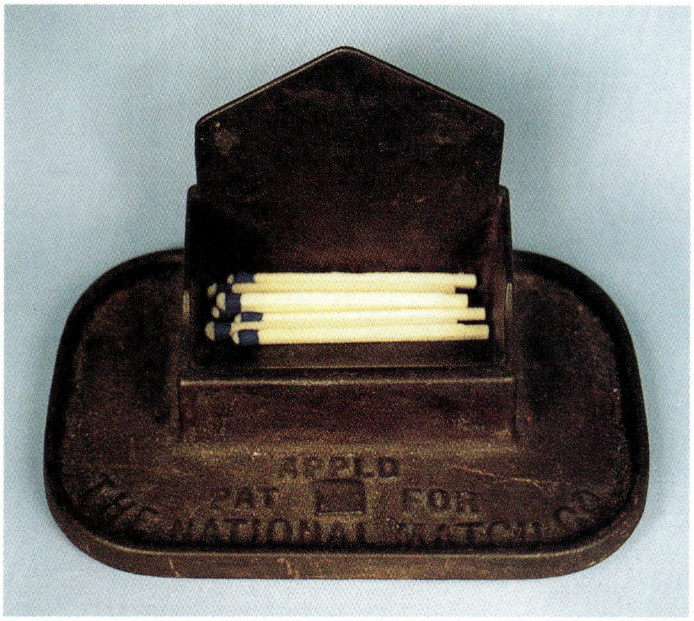

Open match holder which facilitates the removal of one or more matches. The inscription on the lower front reads "PAT APPLD FOR THE NATIONAL MATCH CO." Iron. 3 1/4" x 4 7/8", 2 5/8" tall. *Courtesy of Irontiques*. $75-95.

Claw footed brass match safe with a hinged lid. The rough wood grain design on the top of the box is used to strike the match. Brass. 2 1/2" x 3 3/4". *Courtesy of Irontiques*. $60-85.

Match safe with a hinged lid. There is a rough surface on the top of the lid to function as a striker. Simple design with scrolled feet. Marked under the lid "Bradley & Hubbard West Meriden, Conn. 1808." Iron. American. 3" x 4 3/4", 1 1/4" tall. $95-150.

Match safe on a stepped base. Egyptian theme with hieroglyphics around the sides. The hinged lid has a lion on the top and lifts up to expose the matches inside. There is a ribbed strip on the front of the base to function as a match striker. White metal. $145-175.

Brass match container with a separate lid and a decorative knob on top. The box is covered with a pointed leaf design. The striking surface is under the lid as well as on the bottom of the box. 21/2" x 4 1/4", 2 1/2" tall. $65-95.

Attractive match box with embossed roses on the hinged lid and the front of the container. The striking surface is located under the lid as well as on the bottom of the box. Iron. 2 1/2" x 3 1/2", 1 1/4" tall. $75-125.

Urn on a pedestal match container. The hinged lid is ribbed on the top to provide a striking surface for the match. There is a decorative border around the outside edges. Iron. 3 1/8" x 4 3/4", 2 1/4" tall. $100-150.

Victorian ginger bread style match box on a pedestal base. The top of the hinged lid is corrugated to be used as a striker. The number 12699 is incised under the lid. Iron. 4 1/2" x 5", 2 1/4" tall. $100-175.

A hinged lid match box with a medallion at the top and another medallion on the front. A floral design adds to the decoration. Rosettes on the corners bracket the ribbed striking surface on the lower front of the base. White metal. $95-125.

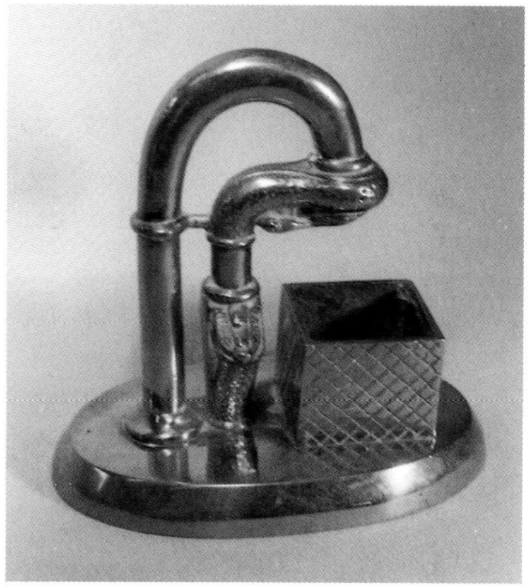

Brass match container made in England. The square box beside the bent pipes is the match holder. The scored sides are used to strike the match. Advertisement reads "Coke Tubes Buckingham Works York." $65-85.

Flat match box covered with a diamond and bar design. The removable lid has a round knob on top with a striking strip underneath. The corrugated pattern around the sides can also be used as a striker. Iron. 4 1/2" x 5". $85-125.

Open view of the flat match box showing the striking surface under the lid.

An ornate floral design covers the surface of this match box. The hinged lid opens to expose a striking surface underneath. The interior is divided into two sections. It stands on four plain feet. White metal. American. 4" x 6 1/2". $125-175.

Open view of floral design match box showing the striking surface under the lid.

Match holder which displays a boy with a chisel and mallet cracking a walnut. The open top of the nut functions as the match container. The striking plate is located in the back. Bronze. $200-300.

Match safe in the form of a devil holding a match aloft. The tiny brass cup in his hand accommodates one match. The back pack can hold a supply of extra matches. The rough surface at his feet is the match striker. White metal. $195-250.

The hunter does not see the rabbit hiding on the lower right. The hollow log holds the matches. The grooved surface on the woodpile in front is the striking pad. White metal. $125-195.

An iron match container in the shape of an urn with handles. The interior is divided in the middle for good matches and for used matches. Iron. Albert D. Judd. American. Patented 1871. $75-125.

A barefoot boy is kneeling beside a hollow tree stump that holds the matches. There is an ax propped on the side and he is holding a tool in his right hand. The composite is supported on four feet. White metal. $125-195.

A nicely detailed match container for the table. A game hunter is tipping his hat while leaning on his gun. A hunter's horn and pouch are around his waist. His game, a rabbit and a bird, are at his feet. The hollow log holds the matches. A leaf design around the base and ornate scroll feet add to the decoration. The striker in the form of a shield is on the front. White metal. 3 1/2" x 4 7/8". $195-295.

Fisherman with an open shirt and rolled up sleeves is holding a basket which functions as the match container. There are fish at the base of his feet. The striking pad is on the right side of the base. White metal. 5 1/4" x 3 1/2", 6" tall. $195-250.

Match container for the table, featuring the figure of a Victorian lady standing beside a basket. The basket functions as the match container. The heart shaped section on the lower front is the striking pad. "AB" is marked in a square on the bottom. White metal. 3 3/4" x 4 1/2", 4" tall. $175-225.

Carved wooden match holder for the table. A squatty little man with a beard and top hat is the whimsical subject of this match holder. The lower front has an opening to hold a small box of matches with the striking surface exposed. The top hat holds loose matches. Probably made in Germany (c. the first quarter of the twentieth century). 2 3/4" x 4 1/8", 4 7/8" tall. $85-145.

A man is sitting on a log beside a basket that holds the matches. A scored striking pad is on the lower front. White metal. $125-175.

Right:
With knees bent, a smile on his face, and a very large tankard in his hand, a man who appears to be tipsey is the theme of this match holder for the table. The basket on his back is the container for matches and has a rough surface for the striker. European (c. the last quarter of the nineteenth century). 2 1/2" x 4", 7" tall. *From the Kurschat collection.* $250-350.

French bronze match holder. A split rail fence and large leaves form the match container. A small hen is on the left side and a rooster is on the right. The rooster is attached to a spindle and can be rotated from a sitting to a pecking position. A ribbed striking plate is on the back side (c.1875). 2 3/4" x 4 1/2", 2 3/8" tall. $600-800.

A bearded street sweeper pushing a broom is the subject of this small match holder. There is a basket beside him to hold matches. Across the front of the base is a corrugated strip for striking matches. Marked "E. G. Zimmerman, In Hanau." Made in Germany (c.1850). 2 1/2" x 3 1/2", 4 3/4" tall. $200-300.

Figure of a man in colonial attire complete with a tricorn hat is the subject of this smoking stand . The square urn on the left is covered with tobacco leaves and holds cigars. The basket in the rear is the match container. An oblong corrugated strip on the right of the base is for striking the matches. The composition stands on four foliated feet. Iron. American (c.1870). 3 5/8" x 6 1/4", 5 3/8" tall. $250-325.

A woman is leaning against a hollow log that is the receptacle for matches. On the front is a basket which functions as the striking plate. Iron. 2 1/2" x 4 1/2", 4" tall. $125-195.

Match holder in the form of a woman selling fish. She has a tray of fish under her right arm and a fish extended in her left hand. She is leaning against a basket that holds the matches. The small basket on the lower front is the match striker. White metal. 4 1/2" x 2 1/2", 4" tall. $125-195.

An open valise is the match containter. The match striker is on the lower front of the bag. White metal. American. $95-145.

Ornate 800 Continental silver match holder with a pierced design around the top and cherubs at play on the sides. Topped with a beaded edge and flowers encircling the base. An oval striking pad is on the right side of the container. 2" x 2 1/2", 2 5/8" tall. $100-150.

A mouse is crouching beside a barrel which holds the matches. The surface of the tray is scored and acts as the match striker. The flanged edge is rippled. Marked "Pairpoint Mfg. Co. Quadruple Plate #1953." American (c.1885). $95-145.

Iron alligator match safe with a hinged lid. Larger than most. 11 3/4" in length. *Courtesy of Irontiques*. $195-250.

A match safe in the form of an alligator. The top is hinged and opens in the back to expose the match compartment. The striker is located inside the alligator's mouth (on the lower jaw). Marked inside is "A. Ziegler." Bronze. 8 1/4" x 3 3/4". $125-195.

Different view of the alligator match safe showing the hinged tail lifted and the space provided underneath.

Geese are taking flight around the sides of this open top match holder. There is a medallion on the base with the inscription "Univ. of Richmond, Va." The rough base is used as the match striker. White metal. American. $95-125.

A fish with a gaping mouth holds the matches. The fish scales are utilized as the match striker. The container rests on a base of seashells and two scroll feet. White metal. 2 1/2" tall. $125-150.

Pretty little oval match box. The striking surface is under the hinged lid and on the bottom of the box. White metal. 2" x 4". $75-100.

Small flat match box. The removable lid features a pug dog resting on top. The striking area is located under the lid. Iron. 3 3/8" x 2", 1 1/2" tall. $95-150.

Close up of match box showing the pug dog on top.

Round match container in the form of a well with a duck perched on the side. The rough surface of the brick is used as the match striker. White metal. 3" tall. $45-75.

Match holder with a removable lid with a dog resting on top. The bin in front holds the burnt matches. The corrugated surfaces on both sides of the box act as the strikers. Iron. 3 1/2" x 5, 3 3/4" tall. $125-175.

Beetle match safe with a hinged lid. Marked "Bradley & Hubbard West Meriden, Conn." Striking area is on the bottom of the beetle. Iron. 2 3/4" x 4". $100-150.

Beetle match safe for the table with a hinged lid. The striker is on the tail and the bottom of the beetle. Bronze. 3 3/4" x 5". $100-150.

An owl is sitting on a branch next to a hollow log that holds the matches. A squirrel is hiding on the left side of the tree branch eating a nut. White metal. Base 3" x 5", 6 1/4" tall. $125-175.

Match safe in the form of a metallic gold striped scarab. The scarab of ancient Egypt was considered a good luck symbol. The shell is hinged in the front and opens to expose the match container. The striker is located underneath the shell. An inscription appears inside in a circle, "G. Hopsess Ct 3200." Iron. 2 3/8" x 4 7/8". *Courtesy of Irontiques*. $495-595.

Match holder with three owls sitting on top of a horse shoe. "GOOD LUCK" is written across the front. The stand on the back allows the match container to sit on a table or be hung on the wall. The striking surface is on the front of the basket. White metal. 6" x 3 3/4". $100-175.

An owl with glass eyes is perched on a tree limb. The hollow tree trunk on the lower left side holds the matches. There is a striking plate on the front of the base. 6 1/2" tall. $125-195.

Iron match safe in the form of a turtle with a hinged lid. Advertisement on the turtle's shell reads "INSURE IN THE OLD CONNECTICUT FIRE HARTFORD." 2 1/2" x 5 5/8", 1 1/4" tall. *Courtesy of Irontiques*. $250-300.

Turtle match safe with a hinged lid. Advertisement reads "GRAND RAPIDS BRASS CO SOUVENIR GRAND RAPIDS MICH." The striking surface is located under the lid. Iron. 2 1/2" x 5". *Courtesy of Irontiques*. $250-300.

Shows the bottom of the Grand Rapids turtle match safe which reads "Mfgrs Refrigerator & Furniture Hdw."

Table match safe in the form of a turtle. The shell is hinged on the side and lifts up to expose the match compartment. The striker is under the lid. White metal. 3" x 5 1/8". $100-175.

Iron match safe in the shape of a frog. The lid is hinged in the back and lifts to expose the match compartment. The rough surface under the lid and on the bottom of the frog are the strikers. Iron. American. 2 3/4" x 3 3/4". $95-150.

Match container with a pair of mice scampering around an old pot. One appears on the top and the other on the lower right. There is a striking strip on the lower front. Bronze. 3 3/4" tall. $200-300.

Match safe for the table in the form of a turtle. The hinged lid opens to expose the match compartment. The match striker is located under the lid. White metal. 2 3/4" x 4 1/2". $95-150.

Match safe in the form of an alligator. Heavy iron with a silver finish. Marked in a circle on the bottom "G. & C. Co. Florida." The striker is located under the hinged lid. 7 3/4" in length. $195-295.

A frog is pulling a shell which holds the matches. The sides of the shell are rough to provide friction to strike the match. White metal. 2" x 4 1/4". $125-175.

Back view of the match container. Notice the beetle getting a free ride on the back of the shell.

Another view of the camel match safe showing the striker under the lid.

A camel in a kneeling position is the subject of this match safe for the table. The saddle on the camel's hump is hinged and lifts to expose the container for the matches. There is a rough strip under the lid to be used as a striker. The front pommel is missing. Marked "Deponirt" under the saddle. White metal. Germany. 2 3/4" x 5". $195-295.

Rectangular match box on a pedestal. The removable lid has a dog with his front feet crossed resting on the top. The square base sits on four foliated feet. The striking surface is on the top of the lid. Brass. 4 1/2" tall. $150-225.

Match holder in the form of a sitting dog with a top hat in his mouth. The rough surface of the base provides a striking surface. White metal with a bronze finish (c.1885). 4 5/8" tall. $175-250.

Dog wistfully looking through a picket fence. The match container is on the back side. There a slot on the front where a piece of emery paper is inserted to be used as the match striker. Iron. 3 1/2" x 4 1/2", 3 3/4" tall. $95-150.

Back view showing the match container and the slot on the front for the emery paper.

A match holder for the table that features a dog sitting beside a basket which holds the matches. The scroll footed base has a striking surface around the edges. Marked on the bottom "Grd 89." White metal. 3 1/2" x 5", 3 3/4" tall. $200-300.

A dove next to a cone shaped basket is the theme of this match holder for the table. There is an oval striking surface at the base of the basket near the bird's tail. White metal. Believed to be from Belgium or France. Base 4" in diameter, 4 3/4" tall. *From the Kurschat collection*. $125-150.

A poodle in a begging pose holding a top hat in his mouth. White metal. American (c.1875). Base 4" in diameter, 4 3/4" tall. *From the Kurschat collection*. $250-300.

Match holder for the table in the form of a fireplace. There is a stand attached to the back to support the upright position. An incised design covers the container. Striker plates are on each side of the fire box. Iron. 4 1/4" x 6". $125-175.

A fence is separating a dog from a ladybug on the other side of the post. Ladybug is there but hard to see. The up turned top hat is used to hold the matches. $95-125.

A blue pressed glass bootee made in a daisy and button pattern. There is a ribbed surface on the bottom of the heel to be used as the striker. Originally made by George Duncan & Sons. The bootee was made off and on from 1938 to 1975 by the Fenton Art Glass Co., Williamstown, West Virginia. Glass bootees have been heavily reproduced. 4 1/4" in length, 2 1/4" tall. $30-45.

A puppy is sitting beside a round cup which functions as the match holder. There is a striking strip on the back side of the cup. On the bottom of the dog is an English registry mark indicating this piece was made Jan. 1, 1882. Above the registry mark is "Deposé." 3" tall. $125-175.

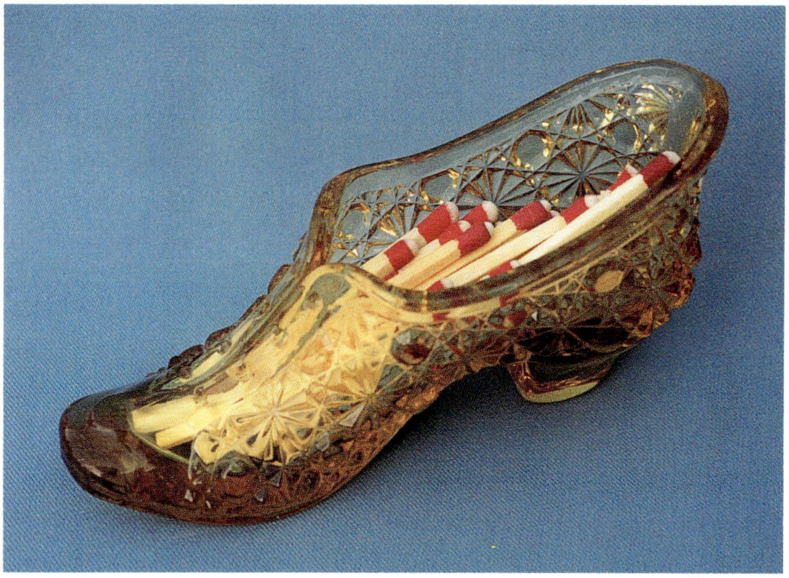

Amber pressed glass lady's slipper in a daisy and button pattern. The rough area on the sole is to be used as the match striker. On the instep of the shoe is inscribed "PATD OCT 19/86." George Duncan & Sons, Pittsburgh, Pennsylvania. 5" in length, 2 1/2" tall. $35-55.

Amethyst pressed glass match container in the shape of a lady's slipper. A ribbed striking surface is located down the back of the slipper. Glass slippers are found in abundance. If there is no striking surface, it was not made for matches. 5" in length. $25-35.

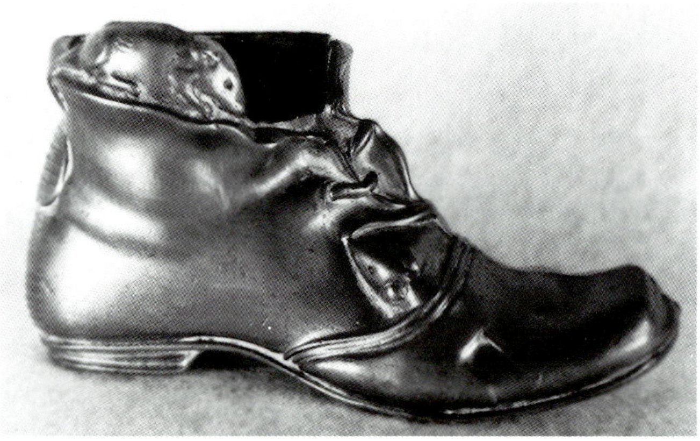

A brass match container in the form of an old shoe. There is a mouse on the top edge and another mouse coming out through the shoe laces. A rough round patch on the heel is the match striker. Marked inside "ENGLAND." $45-75.

A brass match holder in the form of a shoe. There is a mouse sitting on the top edge. Engraved on the side of the shoe is "Ye old lucky match boot." 1 1/4" x 3", 1 5/8" tall. $35-65.

Baby shoe made of white metal. There is a round cup built into the top to hold the matches. On the sole of the shoe is a rough checkered surface to be used as the match striker. 4 1/2" in length, 2 3/8" tall. $35-55.

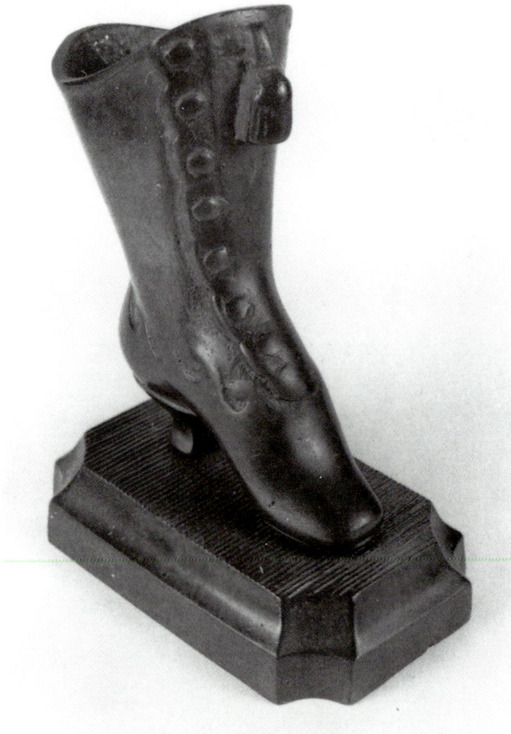

The open top of this lady's boot holds the matches. There are buttons down the side and a decorative tassel at the top. The boot stands on a stacked deck base with a grooved surface on top to be used as a striker. Iron. Base 2 1/4" x 3 1/2", 4 5/8" tall. $95-145.

A lady's boot with a scalloped top and buttons down the side. The open space inside the boot holds the matches. Footed high tiered base. Iron. 3 1/2" x 4 1/4", 5 1/4" tall. *Courtesy of Kossmann's Antiques*. $75-125.

His and hers boots, sitting on a tray with a basket in between. The boots have open tops to hold matches. The woven surface of the basket can be used to strike the matches. The open basket is a handy place to deposit the burnt stubs. Iron. Tray 4" X 5 1/2", 2 3/4" tall. *From the Sanders collection*. $85-110.

Black heavy iron boot with a brown cuff. The open top is used to store the matches. No striker. 3" x 4 3/4". *Courtesy of Irontiques*. $50-65.

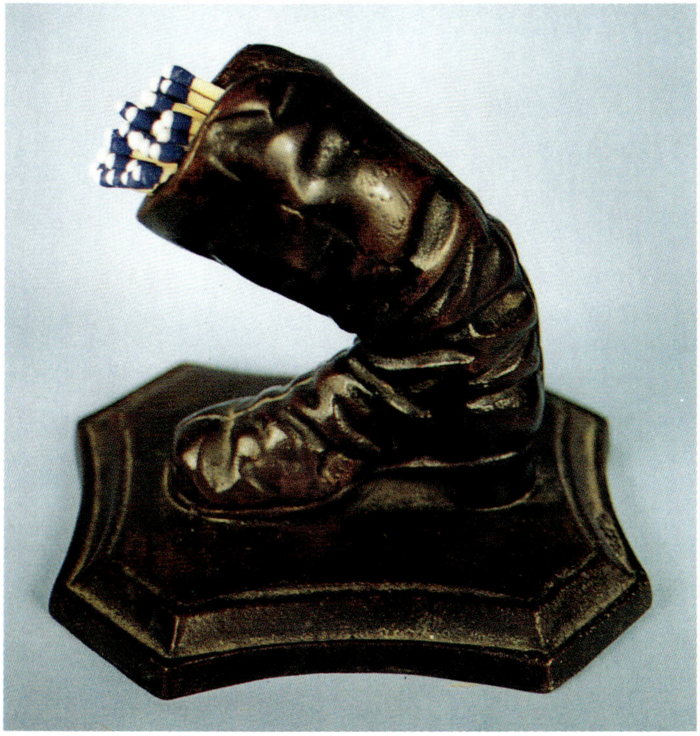

Left:
Match holder for the table in the form of a bent boot. The base is a stacked deck in the shape of a shield. No striker. Iron with a bronze finish. 4 3/8" x 5 1/2", 5" tall. *Courtesy of Irontiques*. $95-145.

A pair of military boots on a platform. The open tops of the boots hold the matches. No striker. Iron. 3 1/4" x 3 3/4", 4" tall. *Courtesy of Irontiques*. $125-145.

Left:
A boot match container. No striker. White metal. American. 2 7/8" tall. $35-55.

A match container in the form of a bent boot. The boot jack has a ribbed surface to be used as the match striker. Stacked deck base in the shape of a shield. Heavy cast iron. American. Base 4 1/2" x 5 3/4", 5" tall. $125-150.

An old worn out shoe with run down heels and brass laces. A round cup inside of the shoe acts as the match receptacle. There is a rough area on the sole to be used as the match striker. White metal. American (c.1890). 1 5/8" x 2 1/4", 4" tall. *From the Kurschat collection*. $100-135.

A shoe with buttons up the side opens on a hinged lid to expose the match compartment. On the sole is a sandy area to act as a match striker. Papier maché. England (c.1840). 1" x 3". *From the Kurschat collection*. $100-125.

A lady's boot with a scalloped top and buttons on the side. The base is a two tiered deck with ribbed striking panels on each side. Heavy cast iron. American. Base 3" x 4 1/2", 5 1/4" tall. $95-125.

Match safe in the shape of a book. The hinged lid is ribbed on the top to form a striking surface. In the middle is embossed "MATCHES." 2 1/4" x 3 1/4". *Courtesy of Irontiques*. $175-225.

Match holder for the table in the form of a rooster. The rear container is for cigars, the square box on the right for matches, and the basket on the left for burnt match stubs. The match striker is located in front on the lower right. Colorful pottery. 5 1/2" x 7", 5 5/8" tall. *From the Sanders collection*. $195-245.

Side view of the match safe showing the spine of the book.

A chest with a rolled blanket on top. The striker is located under the hinged lid. Pottery. Believed to be Austrian. 2 1/4" x 4 1/4", 3" tall. *From the Sanders collection*. $55-75.

A horse with his head lowered out of a paddock. The bucket, with a removable brass liner, is the match holder. Iron. 3" x 5", 5 1/4 tall. *From the Sanders collection*. $275-325.

Reverse side of match holder showing the match striker behind the door. Decorated with a leaf design.

Match container in the form of a shoe. Judging by the sole and heavy laces this shoe was made for rough wear. The delicate bird perched on the top edge of the shoe is a strange addition. 1 3/4" x 4". *Courtesy of Irontiques*. $110-155.

A much plainer version of same style match box. The striking surface is located on the back of the bird. Iron. American. 2 5/8" x 4 1/2", 3 1/4" tall. $125-195.

This charming match holder has been delighting people for over one hundred years. A bird perched on the side of the match box has a double pin in his beak and when his head is pushed down he spears a match and retrieves it. A decorative design appears all around the box. The striking surface is on the bird's back and on the sides of the box. Iron. Made by Sargent & Co., New Haven, Connecticut (c.1885). 2 5/8" x 4 1/2", 3 1/4" tall. $195-250.

Match container in the form of a basket with two hinged lids and a brass handle. Striking areas are between the lids, underneath the lids, and on the bottom of the basket. Iron. American (c.1880). 6 3/4" tall. $125-175.

Hexagon shaped match holder on a pedestal. Panels alternate between pretty floral designs and ribbed striking sections. Hinged lid opens to store matches. White metal. 4 5/8" tall. $95-125.

Bright bronze match stand to hold a small box of matches. This beautiful stand is part of a desk set which includes an inkwell, pen tray, brush stand for cleaning nibs, a seal, and a paperweight. The sides are open to expose the striking surface. Ornate open work design with a bearded man's face in the center. Broad base rests on three bun feet. Believed to be Austrian (c.1885). Base 4" in diameter, 6 1/2" tall. $250-300.

Ornate holder for a small box of matches. The tray holds the burnt stubs. The sides of the stand expose the match striker on the box of matches. American (c.early 1900s). Base 5 1/2" x 5 1/2", 4 3/4" tall. $50-75.

Match container in the shape of a bucket. Applied striking plates are located on either side of the base of the handle. The octagonal tray is a handy place to dispose of the burnt matches. Marked on the bottom "Germany." Tin (c.last quarter of the nineteenth century). Base 4 3/4", 5" tall. $65-95.

This match holder is also a bell. The small lever on the right side, when pushed down, will activate the clapper and ring the bell. The cup on the top with the rough finish holds the matches and also acts as the match striker. This type of match holder was often used in business establishments. $200-295.

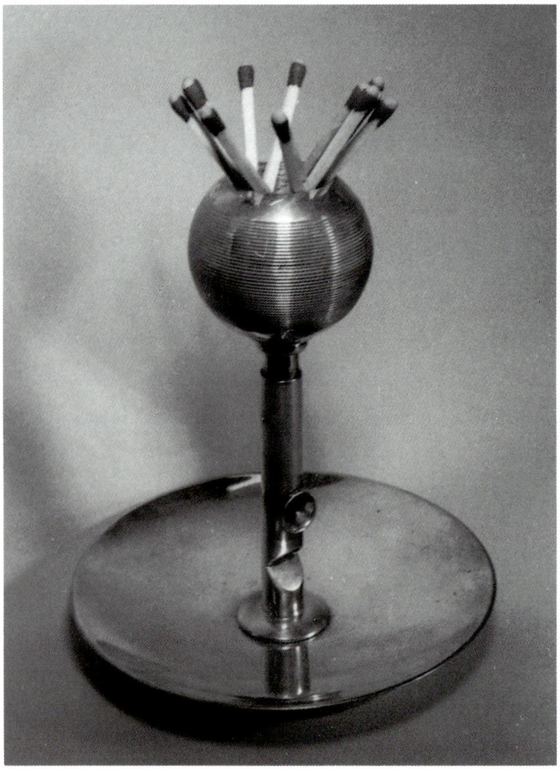

This is a combination match holder and cigar cutter. The ball shaped match container is on a pedestal. The grooved sides serve as the match striker. The cigar cutter is on the lower part of the stem. The under plate is used to collect the burnt match stubs and snipped cigar ends. Silver plate. $175-225.

A match container with two compartments, for either the table or to hang on the wall. The box on the left holds the matches. The burnt matches could be disposed of by inserting them though the slot in the dome on the right. Under the slot is written "SAFETY — 10, 1889." The lid on the dome can be removed for easy emptying. Iron. 2 1/4" x 5 5/8". *From the Sanders collection*. $150-195.

Match container for the table in the form of a light house. The hinged top of the light house opens to store matches. Two boats are nested together. The top one is loose for easy emptying of the burnt stubs. The long ribbed strip on the front of the light house functions as the match striker. Brass. Base 3 3/8" x 5 1/2". *From the Sanders collection*. $175-225.

Lovely porcelain match box with a Mintons hallmark. There is a ribbed match striker under the lid. A turquoise colored band sets off the pretty floral design. England (c.1875). 2" x 3 1/8", 1 1/8" tall. $125-175.

China match box with a Meissen style floral design. Marked "Chas. Ahrenfeldt Carlsbad Altrolau." A ribbed striking surface is located under the lid. 2 1/2" x 3 1/2". $125-175.

Porcelain vase with a ruffled top holds the matches. Painted in shades of pink with purple and red flowers. The striker is on the bottom of the container. Limoges, France. 2 1/2" in diameter, 2 1/4" tall. *From the Kurschat collection*. $35-55.

A porcelain vase for matches. The top edge is scalloped. The sides are covered with scrolls and flowers in rose and gold on a turquoise background. The striker is on the bottom of the container. 2" in diameter, 2 1/4" tall. *From the Kurschat collection*. $35-55.

Celadon match box in an oriental design. When the lid is turned over there is a ribbed surface for striking the match. $45-75.

Porcelain dish with a straight back and a rounded front holds the matches. It is decorated with pretty purple flowers and green leaves. The ribbed striker is across the front. Limoges, France. 2 5/8" x 3 5/8". *From the Kurschat collection*. $45-65.

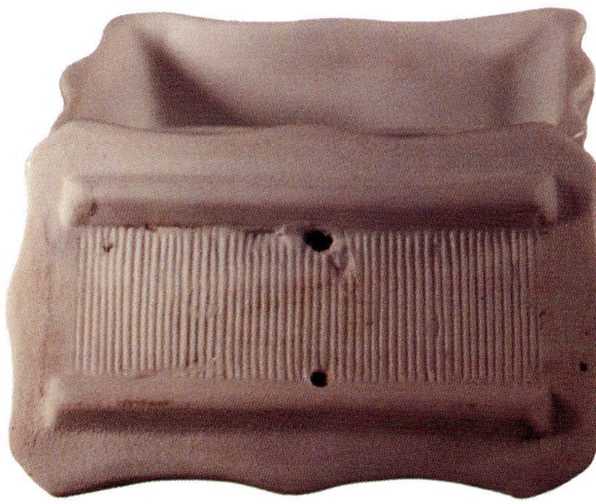

View of the celadon match box that shows the under side of the lid with the ribbed striker exposed.

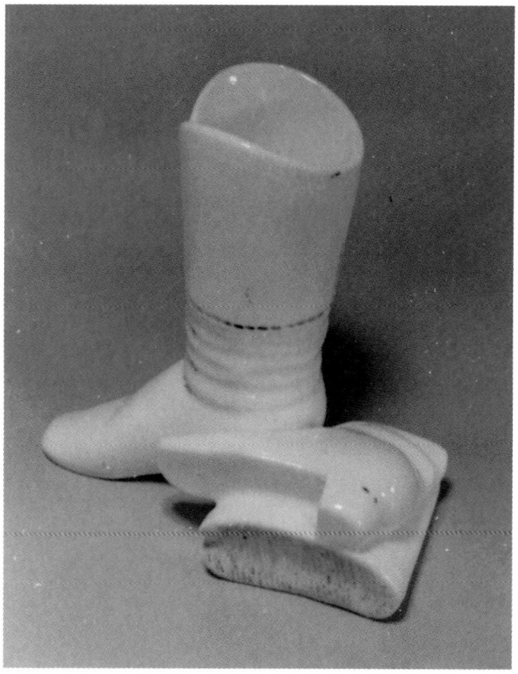

Porcelain match holder in the form of a pair of boots. The folded boot has a ribbed surface in the front for striking the match. $75-125.

Round "Go To Bed" match box. There is a tiny ivory cup on the top to hold just one match. The lid is removable and opens to store the matches. Made in England. $75-95.

Ceramic table top match holder. An Indian baby is depicted tightly laced into a cradle board. There is an open match container on the back with a ribbed area to strike the match. $150-195.

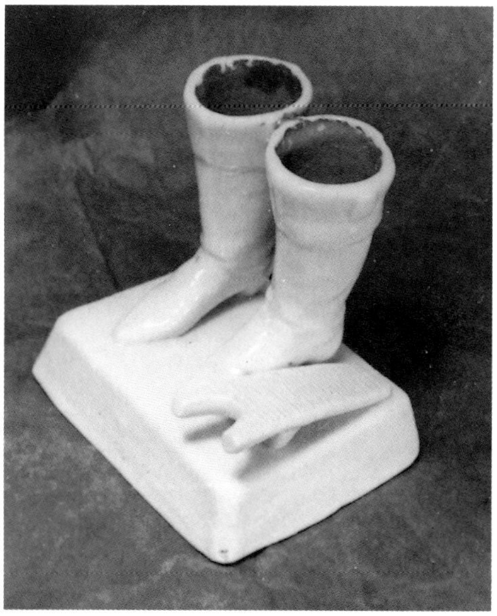

A pair of porcelain boots open at the top to hold matches. The boot jack on the right is ribbed to function as a match striker. Sits on a rectangular base. $75-125.

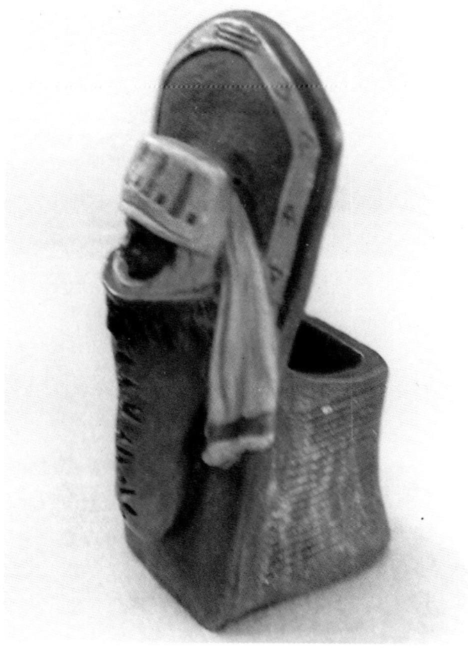

Side view showing the match holder in the back.

A little girl with an apple in her hand is playing keep away with a goose. The match container is in the back. The striker is across the lower front. Made in Germany by Schafer and Vater (c.1915). 3 1/2" x 3 1/2", 3 3/8" tall. $150-200.

Oval "800" Continental silver match box. The lid is hinged and opens to store the matches inside. The striking surface is on the bottom. Base 1 3/8" x 2", 2 1/2" tall. *From the Kurschat collection*. $120-$145.

White ceramic match box with a gold trim. The removable lid reveals a ribbed striker underneath. Probably American (c. late nineteenth century). 2" x 4". *From the Kurschat collection*. $55-75.

This table top match container presents a red bearded man smoking a pipe. With his face in his hands, a bottle between his bare feet, and a smile on his face makes one wonder what is in the bottle. Made in Germany by Schafer and Vater (c.1915). $150-200.

An open view of the ceramic match box showing the striker under the lid.

Back view showing the match container and a mended patch on the back of the shirt for the match striker.

Cone shaped match holder on a pedestal base. The sides of the container are ribbed to be used as the striker. Iron. Base 2 3/8" in diameter, 3 1/2" tall. *Courtesy of Irontiques*. $45-55.

Porcelain match holder for the table. A woman with bare feet has her right hand extended in a begging pose. She is carrying a baby on her shoulders. The match container is in the back and has a rough surface for striking the match. Germany (c.1860). Marked with a shield, a bent arm, and #4179. Conta and Boehm. $195-$250.

Match container and cigar cutter. The round cup holds the matches. Advertising on one side of the cutter reads "SMOKE ADELINA PATTI CIGARS." The other side reads " THE JOS. R. PEEBLE'S SONS CO. DISTRIBUTING AGENTS CINCINNATI, O." There is a woman, partially draped, stretched out on the side of the tray. A long strip on the outside edge of the shell is corrugated to provide friction to strike a match. Iron. 4 3/4" x 6", 5 1/4" tall. *Courtesy of Irontiques*. $375-425.

Back view showing the striking strip for the match.

Porcelain match container. The subject is a bearded man, a peddler perhaps, with a cask on his back and a flask in his hand. There is a bag around his waist and he has bare feet. The match container with a striker is on the back. $150-250.

Side view of the porcelain match container showing the match box with the striking panel in the back.

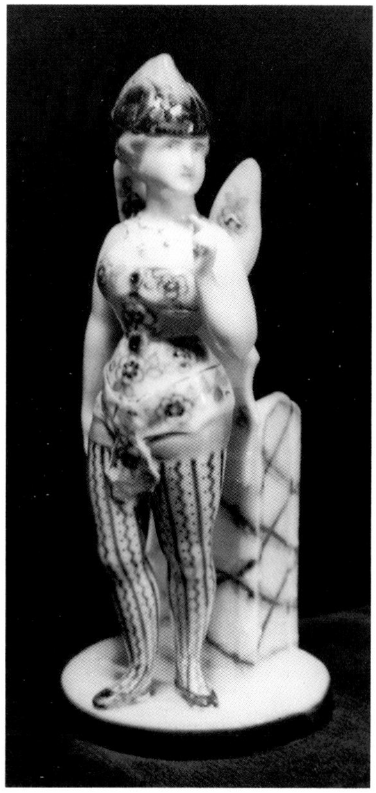

A costumed lady in tights with wings on her back. The match container stands behind her and has a rough surface for the match striker. Marked with a slanted "H" in blue and #409. Porcelain. $175-250.

A shepherd boy eating a piece of bread is the subject of this table top match container. The leaf shaped cup next to him holds the matches. The striking area is on the back side. Bisque. Germany. $125-175.

A boy is standing beside a tulip shaped cup that holds matches. The striker is on the back side of the container. Porcelain. Germany. $125-175.

Round ceramic match container with an open top and a green pointed leaf design. The striking surface is on the bottom. $45-65.

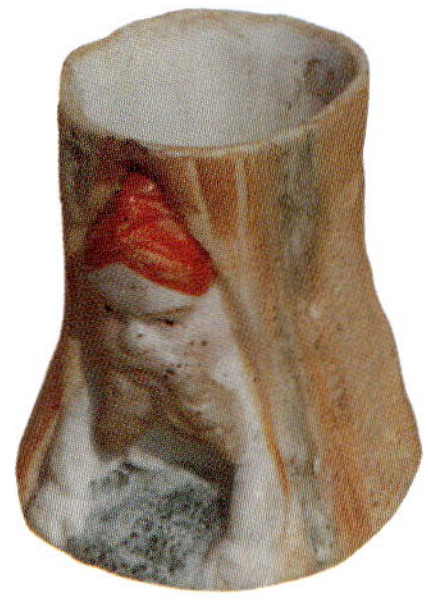

Open top match container in the form of a hollow tree trunk. A woodsman is crawling out of the opening in the front. The striking strip is on the right side. $35-55.

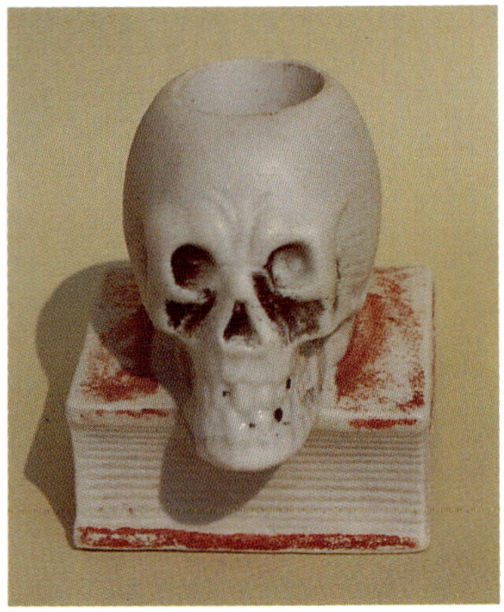

Open top ceramic match container in the form of a skull which sits atop a book. The ribbed pages of the book provide the striking surface. $35-55.

Open top match container colored in dark blue and white with a touch of red. The striking surface is on the back of the holder. The used matches are deposited in the tray. $45-75.

Left:
Porcelain match container presenting a pretty young girl holding a puppy in one hand and an umbrella in the other. The pitcher beside her has an open top to hold the matches. The striker is on the back. Side of the pitcher marked "B. B. 57." Inscribed on the base "Beauty to start." Germany (c.1885). Conta and Boehm. $200-300.

Right:
A ceramic match container depicting a Kate Greenaway type figure of a young girl dressed for winter in a coat and hat. There is a match box in the back with a striker on the side. Impressed on the bottom is a shield and #403 (c. the last quarter of the nineteenth century). $150-225.

Charming ceramic match holder presenting a black boy wearing a pan on his head for a hat and sitting in an old shoe. He doesn't quite fit. His feet are poking out of the open toe of the shoe. $125-175.

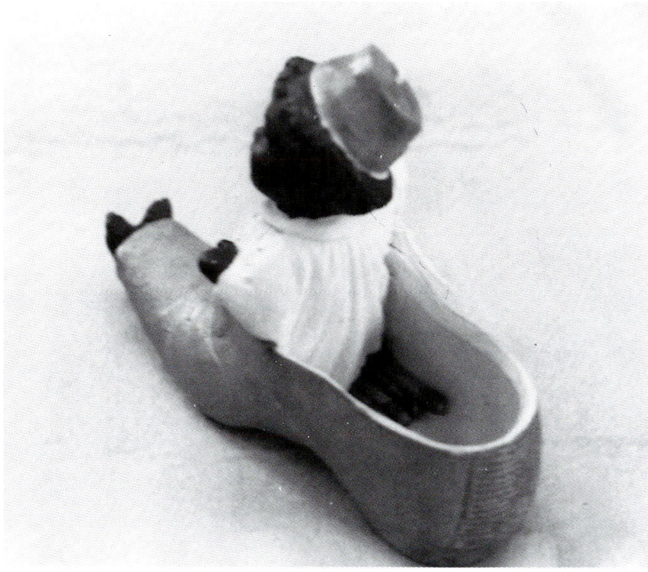

View of the match holder showing the space behind the boy where the matches are stored. The ribbed striking panel is on the back of the heel.

American belleek match container with a floral design around the base. A section of the front is scored to function as the match striker. $50-75.

Ceramic match box with beautiful applied flowers on the top. The removable lid has a match striker underneath. $75-100.

Oval porcelain match box with the portrait of a woman on the removable lid. The striking surface is on the bottom of the box. $75-120.

Left. Porcelain match box with a Dresden style floral painting on the lid. $75-120.
Right. White porcelain oval match box with a gold band around the top. The striker is on the bottom along with a Tiffany hallmark. $120-145.

Left. Oval porcelain match box with an oriental floral design and gold highlights. $75-120.
Right. Oval porcelain match box with a floral design on the top and sides. Limoges, France. $75-120.

Oval ceramic match container with a removable lid. The ribbed bottom of the box provides the match striker. $75-110.

Bottom view of the ceramic match container showing the ribbed striking surface.

Oval pottery match box decorated in a cross and circle design. The removable lid is ribbed underneath to provide the striking surface. $65-95.

Blue, yellow, and white cone shaped match holder on a pedestal. The ribbed cup holds the matches and provides a striking surface. Probably made in England (c.1900). $65-75.

Ceramic cone shaped match holder. Ribbed around the sides of the container to give friction to the match. Advertisement reads "CASSIS QUENOT ME DAILLE D'OR PARIS 1889 TURIN 1900." $50-75.

Cone shaped table match holder advertising "DUBONNET." This style was often used for advertisements. France (c.1910). $50-75.

Round ceramic match cup on a pedestal. A floral design covers the container and the base. The ribbed surface on the top of the base provides the match striker (c.1900). $50-75.

Bulbous shaped crystal match holder with a sterling silver rim. The rough sides are used to strike the match. Hallmark on rim shows the container was made in England in 1898. $150-195.

Round crystal match container with a sterling silver rim. The sides of the glass are ribbed to give friction to the match. Made in England. 2" tall. *Courtesy of "Before My Time" Antiques.* $150-195.

Small match box holder. The openings on the sides expose the striking surface on the match box. The round tray is for ashes and burnt matches. Blue ceramic with gold trim. Made by Hall for the Hilton Hotel. American. 6" in diameter. $30-45.

Heavy crystal paper weight and match container. The rough band running around the top is used to strike the match. Inscribed on the bottom "Webb." Made in England. 3 1/4" tall. This is the larger size (a smaller version was made as well). $200-300.

Stand made to hold a small box of matches. The open sides expose the striking surface on the box of matches. The matches are made by "Bryant and May Co." and printed with "ENGLAND'S GLORY" on the box. $45-65.

This type of match holder was made to hold a small box of matches. Impressed on the bottom is "Fieldings Stoke on Trent #211." The matches are "BRYANT AND MAYS" safety matches. Ceramic. Made in England. $45-65.

Table top match holder designed to hold a small box of matches. A telephone directory forms the match container with a rotary telephone as part of the lid. Copy on the front reads "NATURAL BRIDGE, VA." Brass. American (c.1940). 2 1/4" tall, 2 1/2" wide. *From the Alperton collection*. $40-65.

Victorian dandy with his hat at a jaunty angle. Right hand in his lapel and leaning on a cane, he is a typical "Dapper Dan." At his left side is an open receptacle in the form of a box with a striker on the back side. He is standing on a round wooden base. White metal (c. the last quarter of the nineteenth century). Base 4" in diameter, 10" tall. *From the Alperton collection*. $300-395.

Colonial drummer boy match holder for the table. The background scene depicts a country road leading to a farm house in the distance. The match holder in the back has a ribbed striking surface on either end. Made of pressed or composition material and beautifully decorated. Approximately 4" x 5". *From the Shuman collection*. $300-400.

Match container in the form of a picture frame. Located on the lower front is a single open compartment with a ribbed match striker. Embossed on the reverse side of the frame are the words "ALBUM MATCH SAFE." Iron. 4 3/4" x 7 1/8". *From the Alperton collection*. $195-295.

An opal glass humidor with a painted design of an owl perched on a pine branch against a full moon. Painted in shades of green, rust and brown. The removable metal lid has a barrel on top to hold matches. A Shakespearean verse appears on the back of the humidor. Marked on the bottom with the Handel Ware shield. Meriden, Connecticut (c.1920). 3 1/4" tall. *Courtesy of Andrews Antiques*. $700-1,000.

Brass match container in the form of a boot. The boot jack on the right facilitates the removal of tight fitting boots. The surface of the jack is scored to give friction to strike a match. The base has a pointed leaf design around the edges. Made in Austria. 3" x 3 3/4". *Courtesy of "Before My Time" Antiques*. $225-265.

Two toned brass match safe in the form of a cream pail. The hinged lid opens to store matches inside the container. A grooved striking surface is on the bottom of the pail. Made in England. 1 1/2" x 2". *Courtesy of "Before My Time" Antiques*. $150-195.

Painted white pelican with orange bill and feet sitting on a tree trunk. The hollow log in the back holds the matches and provides a rough surface to act as the striker. The base is light green with gold metallic highlights. White metal. Made by K. & O. Metal Novelties Co., Brooklyn, New York (c.1930). 2 3/8" x 4 1/2", 6" tall. *Courtesy of Andrews Antiques*. $150-225.

Ebony match holder featuring a pelican with glass eyes and a brass cup in his back for matches. The applied metal strip on the back acts as the striker. Made in England. 4" tall. *Courtesy of "Before My Time" Antiques*. $60-85.

Traveling match safe made of tin plate and covered in leather. The pierced striking grid lifts to expose a double compartment for holding matches. Top of the box is stamped "LIGHT." Made in England. Approximately 1" x 2", 2" tall. *Courtesy of "Before My Time" Antiques*. $90-120.

Decorative pottery match holder with the hallmark "Royal D. Eng."(Royal Doulton) on the bottom. The brown glazed band around the top shows a chain and flower design. The rough bottom band functions as the match striker. 4" in diameter. *Courtesy of "Before My Time" Antiques*. $225-255.

View of the traveling match safe showing the open compartments inside for matches.

Round porcelain match holder. The scene on the side depicts a monk reading a newspaper. Made in England. 2 1/4" in diameter. *Courtesy of "Before My Time" Antiques*. $100-150.

Pottery match holder with a Royal Doulton hallmark on the bottom. The brown glazed band around the top is decorated with a club, a diamond, a heart, and a spade. The rough band around the bottom is the match striker. Made in England. 2 3/4" in diameter. *Courtesy of "Before My Time" Antiques*. $120-150.

Side view of the porcelain match holder showing the ribbed striking strip.

Irish bog wood match container. Marked "ERIN" on the back. Decorative three legged bowl covered with shamrocks. 2 1/2" in diameter. *Courtesy of "Before My Time" Antiques*. $60-85.

Table top match safe with two lids that are hinged in the middle and open on each side. There is an opening large enough for the fingers to reach in and extract the matches. An open work handle separates the two compartments. The striking surface is the ribbed section on the lower part of the box. On one lid reads "C. PARKER'S PAT OCT," and the other lid reads "MAY 3, 1870." Iron. 2 1/2" x 3 3/4", 2 3/4" tall. *From the Alperton collection*. $125-150.

Napoleonic soldier with arms folded standing in front of a basket. The basket in the back holds the matches. The striking surface is across the front of the base. Iron. 3 1/2" tall. *From the Alperton collection*. $200-250.

Souvenir match holder for the table. "COTTON STATES EXPOSITION, ATLANTA, GA. 1895. Pat. apld for" on the back. On the front of the bale is "COTTON." A black boy is sitting with his hands on a watermelon and another boy is peeking around the side. Brass. 2 1/8" x 2 3/8", 2 5/8" tall. *Courtesy of Irontiques*. $150-195.

Match holder for the table featuring the figure of Aunt Jemima. A barrel decorated with a bow on the front holds the matches and provides a striking surface. Iron. American. 3 3/4" tall. $75-125.

Table top match holder featuring the figure of a crying child who appears to have fallen and knocked over a flower pot. There is a large single compartment for matches on the back side of the figure with a match striker. Cast iron. American (c. late nineteenth century). 6 1/4" in length, 4 1/2" tall. *From the Alperton collection*. $135-175.

A match container in the shape of a boot. The striking pad is located on the lower front. Bronze. England. Base 1 3/4" x 2 1/2", 3 3/8" tall. $125-150.

A Victorian high heeled shoe with buttons on the side. Iron with an aged bronze finish. American (c. the last quarter of the nineteenth century). Base 3" x 4", 5" tall. *From the Alperton collection*. $65-95.

Table top match holder in the form of a woodpecker perched on a hollow log. The hollow log holds the matches. When pivoted downward, the two sharp prongs on the bird's beak allow it to spear a match and retrieve it. White metal. American (c. the first quarter of the twentieth century). 4 3/4" in length, 2 3/4" tall. *From the Alperton collection*. $75-125.

Table top match safe in the form of a man with a large hat. He is seated backward on a chair and smoking a cigar. The hat atop his head lifts to expose a receptacle for used matches. A second container, in the form of a basket, is on the back of the figure. The corrugated surface of the chair and basket provide the match striker. This figure appears as an exact duplicate of a figure on a toy bank manufactured by the McLaughlin Bros. of Springfield, Mass. White metal. American. Patented 1877. 3 1/4" wide, 5 1/4" tall. *From the Alperton collection*. $295-395.

A camp fire is the theme of this match container. Notched tree branches support a pot hooked to a crossbar. Dirt is mounded to form a pit where wood is placed under the pot. A leaf with a long stem decorates the side. The rough surface of the base may be used as the striker. Quadruple silver plate #2623. Made by James W. Tufts, Boston, Massachusetts (c.1880). *From the Kurschat collection*. $225-275.

Table top match container in the form of a fisherman's creel leaning against a bed of rocks. The rough sides of the basket may be used as the striker. White metal with a round black marble base (c. late nineteenth century). 3 1/2" tall. *From the Kurschat collection*. $135-165.

A whimsical table top match holder. A cat inside a barrel yowls as a monkey on the side pulls the cat's tail through the bunghole. The rough surface of the base or the ribbed rings on the barrel may be used as a striker. Bronze or patinated brass. Believed to be American (c. the third quarter of the nineteenth century). Base 4 1/4" x 5 1/2", 5" tall. *From the Kurschat collection*. $350-450.

A table top match container in the form of a three legged kettle swinging from a chain on a tripod. The inscription on the front reads "MOTHER." Quadruple silver plate. American (c.1880). Spread of tripod 4 3/4", 7 1/2" tall. *From the Kurschat collection*. $135-155.

Table top match container in the form of a burro. The open topped panniers strapped to his back hold the matches. The rough surface of the base and basket may be used as strikers. Vienna Bronze. Base 3" x 4 1/2". *From the Farriester collection*. $400-500.

Match safe in the form of an alligator. The top is hinged on the side and opens to expose three compartments. Marked "Made in England T.H.W. Linton PTR' No 732221." The English design number indicates this item was made in 1927. Heavy iron. 3 3/4" x 9 3/4". *From the Farriester collection*. $150-250.

Iron match safe in the form of a beetle. The lid is hinged at the neck and lifts to expose the match compartment. Inscribed in a circle on the beetle's back is an R with a BC incorporated in the R. The tail is corrugated to provide the striking surface. Patinated brass. 3 1/2" x 4 1/2". *From the Farriester collection*. $175-250.

A match safe for the table in the form of an Egyptian sarcophagus. The top is hinged on the side and opens to store the matches. Inscribed on the inside of the lid is #6968. The striker is under the lid and in the shape of a mummy. Iron. 1 3/8" x 4 3/8". *From the Farriester collection*. $300-400.

Ornate smoking stand on a tray with handles. The rack in the middle is made to hold a cigar. One urn holds the matches while the other is designed to act as a cigar snuffer. To extinguish the cigar, the hot end of the cigar is inverted in the cup. The striker plate appears on the front of the match urn. Brass. 3 5/8" wide, 9 1/8" in length. *Courtesy of Sunny Lenzner Antiques*. $225-300.

Left:
Open view showing the ribbed striking surface under the lid in the shape of a mummy.

Right:
Match container for the table with a gondolier wielding an oar. The match container is in the shape of a scoop. The rough design around the sides of the holder may be used to strike a match. Bronze. 3 1/4" tall. *Courtesy of Sunny Lenzner Antiques*. $225-275.

A Peruvian man with a large cone shaped container on his back to hold matches. Marked "Reyna 925" which, indicates this piece is made of sterling silver. 2 3/8" tall. *Courtesy of Sunny Lenzner Antiques*. $75-95.

Match container for the table featuring a different scene on each of its four sides. Rams' heads appear in the upper corners and the feet are in the shape of hooves. The striking area is the rough edge that runs around the top. White metal. 3 5/8" tall. $95-150.

Wooden smoking stand on a round base. A tobacco container on a pedestal is flanked by two small urns. One may hold good matches while the other may be used for burnt matches. In the front is a cigar cutter in the shape of a house. A small handle with a wooden knob activates a guillotine to cut the end off of the cigar. *Courtesy of Sunny Lenzner Antiques*. $150-200.

Match holder displaying the full figure of a partially nude woman with wings on her head. The arrow shaped shield most likely once held a thermometer. The basket to her side functions as the match container and the rough sides may be used to strike a match. White metal. $125-150.

Brass fly match safe. The wings are hinged to the head and lift to expose a compartment for matches underneath. The fly match holder has been heavily reproduced. 2 3/8" x 3 1/2". *From the Farriester collection.* $150-200.

Open view of the brass fly match safe showing the ribbed striking surface under the wing.

Combination match safe and inkwell. The match container has a separate lid with a striking surface located underneath. The ink stand has a hexagon shaped glass ink bottle with a cone shaped lid. The pen rack in the back holds two pens. Iron with a bronze patina. $295-395.

Combination match safe and inkwell. A glass ink bottle with a separate lid fits in the stand. The pen rack in the back holds three pens. The match box has a separate lid with a ribbed surface underneath for striking the match. American (c.1880). 5 1/4" x 7 1/2", 3 7/8" tall. $295-395.

Smoking stand in the shape of a dog. The dog's head is hinged and opens to reveal a compartment to hold a flint match in a reservoir of lighter fluid. There is a steel striking strip on the side. When the flint match is struck against the steel, a spark is created resulting in a flame. A pipe stand and ash tray are exposed when the hinged door on the dog's back is opened. Underneath the dog is a trap door that opens so the ashes can be removed. Vienna bronze. 6 1/4" in length. *From the Farriester collection*. $150-225.

View of the dog smoking stand showing the pipe stand under the dog's back.

A black match boot on a rectangular base advertising Woonsocket Rubber Co. Inscribed around the base is the message "A MATCH FOR YOU BUT NONE FOR WOONSOCKET BOOTS." Underneath the base is a ribbed striking surface. Iron. American. Base 1 5/8" x 2 5/8". $125-175.

A smoking stand for the table with a hollow log to hold cigars in the back and a smaller log on the left side to hold the matches. The depression in the lower front is for the burnt stubs. The full figure of a ram leaping over a log is displayed on the front. The overall rough surface provides ample striking surfaces. Engraved on the bottom "CS 1658." Terra cotta. Oval base 3 3/4" x 5", 4 1/2" tall. $125-175.

A beautiful Vienna bronze combination inkwell and match holder. A barefoot boy is holding a hunting gun in one hand and a match in the other. The dog on the left side appears to be interested in the match. The inkwell with a hinged lid is on the right side, and the match container is on the left. A hunter's pouch is on the lid of the inkwell (c.1865). Oval base 3 1/2" x 5 1/4", 4" tall. *From the Crabb/Nomann collection*. $800-1,200.

A French bronze combination match holder and inkwell. The figure of a boy is standing in the middle of the composition. The inkwell with a hinged lid is on the right side, and a hollow log that holds the matches is on the left. A hunter's pouch is on the lid of the inkwell (c.1865). Oval base 3 1/2" x 5 1/4", 4" tall. *From the Crabb/Nomann collection*. $800-1,200.

Another view of the above match holder, inkwell combination showing the open inkwell. There is a small hole under the boy's right hand to hold a single match.

A tin frame fits over a cardboard box that holds long fireplace matches. On the front is an attractive scene showing two men fishing. On the other side of the river is a large colonial house. On the sides of the box are oval striking surfaces. 4 5/8" x 7 1/2", 11 3/4" tall. The matches are 11" in length. *Courtesy of Eve Naslund*. $125-150.

A smoking stand for cigars, a pipe, and matches. Three baby chicks have just emerged from their shells. One still has the top of the broken egg shell on his back. The match holder is an egg with the top broken off. Marked "L. & C.12." Terracotta. Base 6 3/4" in diameter, 5 1/2" tall. *From the Crabb/Nomann collection*. $150-225.

An ornate match stand for the table. The match container has a cut out pattern and fits over a copper box. There are striking strips on each side. Features include curled leaves on a twisted stem and three leaves that are the legs. The tips of the leaves are curled up to form feet. Decorated with copper flowers, bands, and brads. Iron. 6 1/4" tall. $100-150.

An Arts and Crafts match holder for the table in the form of a whimsical animal. The cup holds the matches and there is a long ribbed striking strip down the front. Bronze. American. 3 1/4" tall. *Courtesy of Linda Fisher and Andreas Kazos*. $300-500.

A smoking stand for the table. An American buffalo is standing between a hollow log on the left that holds cigars and one on the right that holds matches. Tin. American. Odd shaped base 5 1/2" x 9 5/8", 4" tall. *Courtesy of "Treasure of the Sierra Madre."* $125-195.

A table top match holder. A boy dressed rather elegantly is standing next to a white chicken. The match container, behind the boy, has a ribbed square block to function as a match striker. Colorful bisque ware. European (c. last quarter of the nineteenth century). 4 1/2" tall. *From the Kurschat collection*. $75-110.

Back side of the table top match holder on the left, showing the location of the match container and striker.

A small cup with a large handle is the container for matches. The sides of the cup are rough to provide friction for striking the match. White metal. 1 1/4" in diameter, 1 5/8" tall. *From the Kurschat collection*. $25-45.

A richly colored and glazed majolica match holder for the table. A comical character is standing with one hand on a large shoe that holds the matches. The back side of the shoe has a ribbed block for a striker. Believed to be French (c.1900). 4 7/8" tall. *From the Kurschat collection*. $110-175.

Back side of the match holder on the left, showing the ribbed striking block on the shoe.

A small barefoot boy is holding a large boot that holds the matches. The figure stands on a marble base. Bronze. European. Base 3 3/4" in diameter, 4 1/2" tall. *From the Gersten collection*. $250-350.

The feathered headdress of an Indian is open at the top to hold matches. Heavy bronze. American. 2 3/8" x 3". *From the Gersten collection*. $125-150.

The head of an Indian with three feathers in his flowing hair is the subject of this container for matches. Brass. American. 1 1/2" x 2 3/8". *From the Gersten collection*. $50-75.

A cat is sitting in a regal pose beside a basket that holds the matches. The match striker is located on the base. The basket can be used as the striker as well. Hand forged nuts hold the figure of the cat and the basket to the base. Brass (c. the last quarter of the nineteenth century). Base 3 1/2" in diameter, 3 3/4" tall. *From the Gersten collection*. $250-350.

A sterling silver match box holder in the Repoussé pattern by "S. Kirk & Son Inc." A very small match box fits in the holder. The striker on the match box is exposed on the sides. Displayed in a 1932 Kirk catalogue. 1 1/4" x 1 3/4". $50-60.

A whimsical match cup for the table. A pair of monkeys resting on their tails form the handles of the cup. Brass. The cup is 2 1/2" in diameter. *From the Gersten collection*. $50-85.

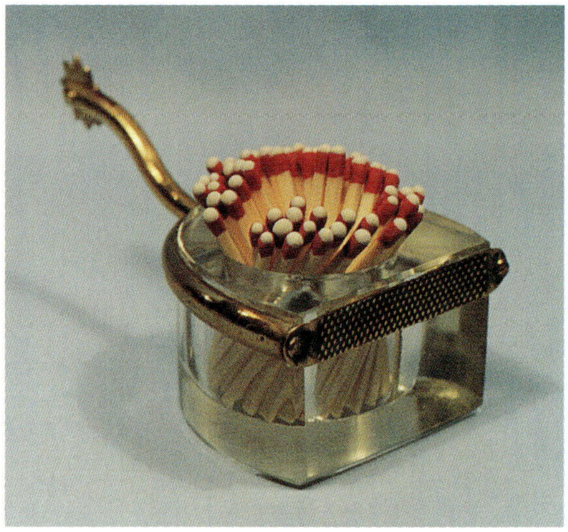

Glass match holder for the table with a brass frame in the shape of a spur. The front of the spur is scored for striking the match. 2 1/4" x 4". *Courtesy of Mizrahi Antiques*. $150-175.

Inside view of the match box showing the oval striking pad under the lid.

Beautifully decorated match box. The hinged lid displays a cherub on top. A medallion decorates the front and there is a floral design around the sides. Stands on four scroll feet. The striker is under the lid. Heavy white metal. 4 1/8" x 6", 2 3/4" tall. $200-275.

Unusual table top match holder in the shape of a pumpkin. The hinged lid opens to store the matches inside. There is a rough strip along the front of the base to be used as a striker. Iron (c. the last quarter of the nineteenth century). Base 4 1/2" x 4 1/2". *From the Alperton collection*.$250-300.

Colorful ceramic smoking stand in the form of a cooper working on a barrel. The large barrel is to hold cigars and the small barrel for matches. The broad corrugated strip across the lower front is the match striker. European. 6" x 6", 5 3/8" tall. $250-350.

Brightly colored smoking stand for the table. A gentleman, wearing knee breeches and a large hat, is sitting on a stool. The large barrel is for cigars, the small one is for the burnt stubs, and the mug holds the matches. White metal. 3 3/4" x 7", 6" tall. *From the Alperton collection*. $150-250.

A bronze turtle holds a small box of matches. When the turtle's head is pressed down the match box is pushed out the back. The sides are open to expose the striking surfaces on the match box. The bottom of the turtle is inscribed, "Barkdull." 3" x 4". *Courtesy of Trail's End Antiques*. $55-85.

A gas stove forms a box for holding matches. The lid with three burners opens by pressing two levers in the back. On the front of the oven door is "Savage." In a circle on the back of the stove is the message "AN ORIGINAL SAVAGE CALIFORNIA DESIGN PAT PENDING." Brass. American. 4 1/4" tall. $125-175.

A round brass basket holds the matches. The bird on the outside edge sits on a spindle and can be freely moved around the base. The scored surface of the round base and basket can be used to strike the match. Brass. Made in England. Base 3" in diameter. *Courtesy of "Before My Time" Antiques*. $135-175.

Paper matches were invented in 1892 by Joshua Pusey, a Philadelphia patent attorney. They are small and easy to carry in the pocket. More paper matches are used today than any other kind. A great many of them bear advertisements and are provided free to customers.

A black and white terrier with a red collar corners a cat on the top of a shoe. The striker appears as a ribbed patch on the heel of the shoe. Ceramic. Made in Japan. 3 5/8" in length, 3 3/4" tall. $30-45.

A top hatted owl is standing beside an easel which is designed to hold a book of matches. The disk on the front of the owl reads "Summit of Pikes Peak Alt 14106 ft Colorado." White metal. Owl 1 1/4" x 2", stand 3" x 4". *From the Farriester collection*. $125-200.

A green marbleized Akro glass match stand. The round tin platform at the top holds a paper cylinder of matches. The knob on top unscrews and a refill of matches can be added when needed. The spindle is spring loaded. When the match platform is pressed down, it frees the matches which can then be torn off and lit on the rough surface of the top. Around the bottom of the base is written "Hamilton Match Co. Cincinnati, Ohio. Patented Apr. 29, 1935 & Pats. Pend. U.S. Pat. Off." Manufactured by the Agate Co., Clarksburg, West Virginia. 5 1/4" in diameter, 6" tall. $60-90.

Cylinder of paper matches used to refill the match stand.

"Go To Bed" Match Safes

The legend behind these small match containers is that they were created to burn just long enough to allow a person to jump into bed after extinguishing the candle on the mantle. Since the small flame would burn out quickly, these containers were considered to be safer near bedding than larger candles. For whatever purpose these adorable little figures were made, they are intriguing in their design and highly collectible. They are usually quite small and some have a head that is hinged and opens to store spare matches inside the body. A tiny hole, or cup, is also incorporated into the design to hold one match. The striking surface may be part of the design, for example the bark of a tree, or it may appear as any rough surface. Most were made in England or Germany, circa the last half of the nineteenth century.

A dancing bear with glass eyes is the subject of this "Go To Bed" match safe. The bear is muzzled and has a chain around his neck fastened to a post. There is a small hole in the top of the post to hold a match. The head is hinged in the back and opens to store spare matches inside the bear. Brass. 2 7/8" tall. *From the Farriester collection*. $300-400.

Small "Go To Bed" match container in the form of a knight. The small cup in his hand holds one match that creates a torch when lit. The basket on the side holds the matches and provides a rough striking surface. A leafy design decorates the base. Iron. $200-300.

Left:
"Go to Bed" match safe. A dog with glass eyes is sitting in a begging pose attached to a post. A hole in the top of the post holds the match. The head is hinged in the back of the neck and opens to store spare matches. Iron. Base 1 3/8" x 2", 3 3/4" tall. *From the Farriester collection*. $300-400.

"Go To Bed" match safe in the form of a Napoleonic soldier. He has his left arm behind his back and is holding a rifle in his right hand. The hole for the bayonet holds one match. The head is hinged in the back and opens to store the spare matches in the body. The striker is located in the back of the figure. 3 1/8" tall. *From the Farriester collection*. $225-300.

"Go To Bed" match safe with a colorful picture of a castle on the top. There is a tiny hole on the side of the removable lid to hold one match. The body of the cylinder is open fret work and can hold a supply of matches. Brass. 2 1/8" in diameter, 2 1/2" tall. *From the Farriester collection*. $200-300.

Right:
A round brass "Go To Bed" match safe with a red crackle finish. The tiny cup in the top of the cylinder holds the match. The top is removable to store the rest of the matches in the body of the container. On the bottom is written 'SAFETY BOX 50 PATENT VESTA LIGHT." England. 2" tall. *Courtesy of "Before My Time" Antiques*. $75-95.

A "Go To Bed" match safe in the form of a knight holding a sword in his left hand and a torch aloft to his right. The tiny hole in the torch holds a single match and the basket on the side holds the extras. The scalloped base and basket have a rough surface to strike the match. Marked on the bottom "VERLAG BEI E. G. ZIMMERMANN HANAU." Iron. Germany (c.1855). 2 1/2" x 3 3/4", 4 1/8" tall. $200-300.

"Go To Bed " match container with a soldier ready to fire a mortar. There is a tiny brass cup in his hand to accommodate one match. The open mouth of the mortar is used to store the extra matches. $200-300.

Close up view of the soldier match container showing the English registry mark which indicates this piece was made in England, on July 30, 1850.

An owl sitting on a twig is the theme of this "Go To Bed" match safe. The head opens in the back of the neck and the matches are stored in the body. There is a small cup on the owl's head to hold one match. The striking surface is around the base. Base 2 5/8" x 2 7/8", 3 5/8" tall. *From the Farriester collection*. $200-300.

Circus bears performing a balancing act form this delightful "Go To Bed" match holder. The top bear stands on the feet of the bottom bear while balancing a torch in his mouth which holds the match. The scored base may be used as a striker. There is no container for extra matches. Base 2" x 2 3/8", 4 1/2" tall. *From the Farriester collection*. $250-350.

"Go To Bed" match safe in the shape of a small pig with a corkscrew tail and a hole in his snout to hold a match. The pig's head is hinged at the bottom of the neck and opens to hold the spare matches. The striking surface is on the bottom of the pig. Brass. 1 3/4" in length. *From the Farriester collection*. $195-250.

View of the pig match safe showing its hinged head.

"Go to Bed" match safe in the form of a pig sitting on his hams clutching a football. A hole in his snout accommodates the match. The head is hinged and opens to store the spare matches. The striker is located on the bottom. Brass. 2" tall. *From the Farriester collection*. $195-250.

Small "Go To Bed" match container in the shape of a mortar. The tiny hole in front is for burnt match and the one in the back is for the good match. The ribbed block in front of the mortar is the striker. Iron. 3 3/4" x 3 1/4", 3" tall. $195-225.

"Go To Bed" match holder in the form of a knight with one hand on his hip and the other holding up a match. A basket to hold matches is located in the back and has rough sides to be used as a striker. The base is decorated with leaves and stands on four foliated feet. 2" x 2 3/8", 4" tall. *From the Farriester collection*. $200-300.

Olive wood "Go To Bed" match safe with a scenic print on the front. The tiny ivory cup in the top holds one match. The round lid is removable and extra matches may be stored inside. A piece of sand paper under the pedestal base may be used to strike the match. Made in England. 5 1/2" tall. *Courtesy of "Before My Time" Antiques*. $125-175.

"Go to Bed" match safe in the form of an owl with glass eyes. The head is hinged at the neck and opens to store matches inside. A tiny hole on the right side of the branch holds a match. Gilded bronze (c.1890). 2 1/2" tall. *Courtesy of "Before My Time" Antiques*. $275-325.

Side view of the match safe showing the hinged head of the owl.

Figure of a woman dressed in a nightgown and a mob cap. The cup she carries on a tray holds the match. The size of the match to fit in the cup is rather large for a "Go To Bed", a small birthday size candle would fit. Her head is hinged and lifts so spare matches can be stored inside the body. The criss cross design around the base is the match striker. Iron. Base 2" in diameter, 3 3/4" tall. *From the Farriester collection*. $300-400.

Rare flow blue porcelain "Go To Bed" match safe. The lid is removable and the spare matches can be stored inside. The rough bottom of the container provides the striker for the match. England. 3 1/2" tall. *Courtesy of Sunny Lenzner Antiques*. $400-500.

Round brass "Go To Bed" match safe. The message on the bottom of the container around the edges reads "G + B Vesta Lights 100 Patent Prince Albert's Safety Box." The lid lifts off and the extra matches may be stored inside the cylinder. Made in England. 2 1/4" tall. *Courtesy of Sunny Lenzner Antiques*. $175-200.

Egyptian hieroglyphics decorate this "Go To Bed" match safe. A small cup on top holds the match. The front is hinged and lifts to expose a storage compartment for extra matches. The inside compartment also reveals another scene depicting early Egyptian life. Under the lid is the small striking plate. Stamped brass. 2 3/8" x 4". *From the Farriester collection*. $200-300.

A whimsical French bronze "Go To Bed" match holder. A pair of foxes march across an oval alabaster base. The fox in front hauls a cup on his back to hold matches. The fox in the rear, wearing a night cap and carrying a cane, lifts a stein that holds one small match. The oval striking pad in the front is ribbed to supply friction for a match. 2 1/4" x 4", 3 1/4" tall. *Courtesy of Mizrahi Antiques*. $395-425.

A bearded and mustached soldier is the subject of this "Go To Bed" match safe. It is hinged in the back and opens to store extra matches in the body. The gun barrel holds a match. Brass. 3 1/2" tall. *Courtesy of Mizrahi Antiques*. $375-425.

Beautiful bell shaped "Go To Bed" light. The hinged lid opens to store matches in the body of the container. On top is a candle holder and two small holes. A match fits in one hole; the other hole is for the hook on the chained candle snuffer. The top of the finger ring is scored for striking the match. Beautifully impressed leaf decoration. Brass. Made by the Harcourt Brothers in Birmingham, England (c.1852). Base 3" in diameter, 2 1/2" tall. *Courtesy of Eve Naslund*. $295-350.

Back view of the soldier match safe showing the backpack with a scored surface to strike the match.

A brass "Go To Bed" match holder. A man in a robe and night cap appears to be groping his way down a stairway with a torch held high to light his way. A tiny hole in the torch holds the match. The stairs are covered with an oriental carpet runner. The striking surface is on the steps. There is no container for extra matches. European (c.1880). 5 3/4" tall. *Courtesy of Mizrahi Antiques*. $425-495.

A "Go To Bed" match safe in the form of a man smoking a pipe. The peaked hat is hinged in the back and opens to store the matches inside the body. The bowl of the pipe may hold a single match. The scored surface around the base may be used as the match striker. Iron. Made by E. G. Zimmerman in Hanau, Germany (c.1850). 4 1/2" tall. *From the Alperton collection*. $250-300.

A crowing rooster with red glass eyes is the subject of this "Go To bed" match safe. His head is hinged in the back and opens to store extra matches inside. The small hole on the base holds a single match. The base is rough and can be used to strike the match. Brass with a bronze patina. 4 1/4" tall. *Courtesy of Mizrahi Antiques*. $375-425.

Beautiful "Go To Bed" match safe with the Queen of Hearts standing on top of a round match container. On her head is a small cup to hold a single match. The container opens for extra matches to be stored inside. The square base has jewels set on all four corners. Bronze. French. 4" tall. *Courtesy of Mizrahi Antiques*. $475-525.

Chapter Four

Match Holders For The Wall

Primarily, match containers were hung on the wall for safety. The highly flammable matches were kept out of reach from children and were protected from rodents. They could be placed where they were needed most, by the cook stove, fireplace, gas light, or anywhere in the house. Their fixed location made it easier to find them in a dark room.

A roofed well functions as the match container for this very attractive match holder. There is a split rail fence in the background with a mouse on the lower left side and another on the lower right side. Grapes and vines add to the decoration. A match can be struck on any of the rough surfaces. White metal. 4" x 7 1/2". $200-250.

Full figure of a man laboring under a load on his back. The barrel is the container for the matches. The background design is a brick wall with grapes and vines. The striker is located under the platform. White metal. 2 3/4" x 8 3/4". $250-300.

An old fashioned telephone at the top of the match holder with "RING UP" embossed below. The advertisement on the lid of the match container reads "DYEING & CLEANING CO. LUNGATROS." On the back is #50. Iron. American. 3 1/4" x 6 1/2". *Courtesy of Irontiques*. $300-450.

Cast iron match container. A barrel for holding the matches is on a rough board fence. A rectangular striking pad is on the right side. American. 4" x 7". *Courtesy of Irontiques*. $100-150.

A match holder in the form of Lucifer captured in a basket. His head, hands, and feet protrude from the front of the container. The top of the basket is open to hold the matches. On the sides of the basket are ribbed striking strips. White metal. 3 1/2" x 6 1/2. $250-300.

Single open top container with a bearded and horned devil's head. Grapes and leaves form the open work background. American (c.1880). This match holder has been heavily reproduced. $95-150.

This match holder is designed to hang on the wall or sit on a table. The message on the hinged lid reads "SELF CLOSING FOR MATCHES & C. PATENTED DEC. 20 1864 D.M. & CO. NEW HAVEN." Iron. American. 2 3/8" x 3 3/4". *Courtesy of Trail's End Antiques.* $85-125.

A plain iron match holder for the wall or table. The checkered design on the hinged lid forms the striking pad. Iron. American (c.1880). $65-95.

Octagonal match holder for the wall. Embossed faces of cherubs cover the background. A baby shoe with wire laces holds the matches. The striking strip is on the lower front. Sheet brass. American. 5 1/2" x 5 5/8". *Courtesy of Trail's End Antiques.* $65-95.

Single compartment match safe for the table or the wall. Message on the top reads "SELF CLOSING FOR MATCHES & C. PATENTED. DEC. 20, 1864 M. W. & CO." Iron. American. Top 3" x 5 1/4". $85-110.

A lacy match holder with a single open receptacle. Graceful open work design appears on the match bin and back plate. The striking surface is on both ends of the container. Iron. American. 3 1/4" x 3 1/4". $75-125.

A single compartment match box with a hinged lid. Designed to hang on the wall or sit on the table. A scroll design decorates the top. American (c.1880). $65-95.

Wall hanging match holder with ornate lattice work and three arches in the background. A single open container with a leaf and floral design on the front. There are striking areas on both sides of the container. Iron. American (c.1880). 3 1/2" x 3 3/4". $100-150.

A match container for the wall with an open work back plate. Flowers cover the open match bin. Striking surfaces are located on both ends and underneath the match compartment. Iron. American. 3 1/4" x 3 1/4". $75-125.

Wall hanging match holder with a single open receptacle. The back plate is an open work design in the shape of a flower and leaves. The match bin has a floral design on the front. Striking surfaces are located on both ends and on the bottom of the match compartment. Iron. American. $75-125.

This match holder has a decorative open work back plate. The single match receptacle has a hinged lid and is covered with a floral design. The match striker is located on the bottom of the container. Iron. American (c.1880). 4" x 6 1/2". $125-150.

Wall hanging cast iron match holder with a single open receptacle. There are striking surfaces on both ends and on the bottom of the container. The back plate has an ornate lattice work design, and there is a leaf and flower decoration on the match bin. American (c.1880). *From the Alperton collection*. $100-150.

Match container for the wall with a single open compartment. Strikers are on both sides and under the receptacle. Also made in a two tiered design. This match holder has been heavily reproduced. Iron. American. 4" x 4 1/2". $75-125.

Grape vines and leaves decorate this match holder for the wall. There is a single open match container with striking surfaces on the ends. Iron. American. 3 1/2" x 5 3/8". $75-125.

Double match container with an open work back plate. There are strikers on both sides as well as the bottom of each match bin. This style was also made with a single match compartment. Iron. American. 4" x 6 1/2". $125-175.

Pretty little match safe with an elaborate open work background. A heavily decorated single match compartment with a hinged lid. The striking plate is on the lower front as well as the bottom of the match bin. This is a smaller version of the match safe at right. Iron. American. 2 7/8" x 6 5/8". $125-175.

Single match holder with a hinged lid. Ornate open work back plate. Strikers appears on the lower front as a long scored strip and on the bottom of the box. Iron. American. Patented 1875. 3 3/4" x 7 1/2". $125-175.

Left:
Cast iron match safe for the wall. The single compartment has a hinged lid. There are strikers on the sides as well as under the receptacle. An ornate design covers the entire subject. 4 3/4" wide, 6" tall. *From the Alperton collection*. $75-125.

Right:
Match container for the wall with a lattice work background and two open top receptacles. The hanging hole is in the shape of a tulip. The top match holder has a large opening so a thumb and finger can be inserted to extract the matches. The lower bin is open to receive the burnt matches and has "IDEAL" embossed on it. This is probably an advertisement for a stove company. A criss cross pattern on a plate across the lower front is for striking the matches. Iron. American (c.1900). 2 7/8" x 6". *From the Kurschat collection*. $85-125.

Single urn match holder for the wall with an open work background and striking bars on both sides. This is a common match holder which comes in many variations: single or double urns, with or without an advertisement. Marked on the back "Pat'd-JAN 15 1867." Cast iron. American. 6 1/4" x 3 3/8". *From the Alperton collection*. $75-100.

Double match holder for the wall. Two urns against an open work background. One container holds the fresh matches while the other holds the used. The striker is the strip in the middle of the lower front. Iron. Made by J. E. Stevens & Co. in a design patented by Russel Frisby. American. Patented 1867. 5 x 7 1/2". $95-125.

Single urn match holder for the wall. The oval on the front could display an advertisement or a painted scene. Striking strips are located on either side of the match container. Marked on the back "Pat'd Jan. 15, 1867." Iron. American. 3 3/8" x 6 1/4". $75-100.

Match holder to hang on the wall. A curious cat is clinging to the side of a three legged kettle hanging on a hook. A rough board fence forms the background. The wash board has a corrugated surface for striking the match. White metal. Probably American. $175-250.

Pretty French match holder to hang on the wall. There is a figure on each side holding torches. The top features a cherub sitting with crossed feet. There are flowers trailing down the sides. The pocket for holding the matches has a flower on the front. The screen in the middle is for scratching the match. Marked on the back "42 A.B. Depose." Marked on the front "Brevete S.G.D.E." White metal. $150-195.

Wall match holder with an advertisement for the Michigan Stove Co. Claims they "ARE THE BEST." Iron with a single container. American. *From the Alperton collection*. $125-150.

Above right:
Eastlake style double match holder for the wall. The top compartment has a hinged lid to safely store the matches and the bottom container is used to collect the match stubs. Covered in an ornate design. There are match strikers located under both receptacles as well as on the sides. Identical to the match holder below, except the mid section is pierced rather than solid. Iron. American. 3 1/2" x 7". $125-175.

Right:
Double match holder for the wall. The top compartment has a hinged lid and is used to store the matches. The open bin on the bottom is used for the burnt stubs. Scrolls and flowers cover the entire subject. This match holder is basically the same as match holder above, but with a solid background rather than a pierced. American. 3 1/2" x 7". $125-175.

A simple, but elegantly, designed match holder for the wall. Striker panels are located on both sides of the open match container. The panel in front displays a small leaf decoration. Iron. American. 4 1/2" x 8 5/8". $125-195.

Single match holder with a hinged lid. Decorative design features a pair of birds sitting in the top branches of a tree. Another bird is on the lower front among the leaves. The striking area is under the box and on the sides. Iron. American. $75-125.

Lovely match holder in the shape of a bench to hang on the wall or stand on the table. A single hinged compartment with a striker across the front. This version is made of iron. It was also made in brass with or without a lid. American. Design patented by Albert D. Judd, 1871. $125-195.

Wall hanging match holder with two open receptacles. Highly decorated with hunting symbols such as a stag's head, a rabbit, a pheasant, a hunter's horn, and a gun. This very popular match container has several design variations as well as three patent dates: September 14, 1868; May 3, 1870; and September 27, 1870. Iron. 5" x 11". *From the Alperton collection*. $175-225.

A hunter's two pocket pouch is the theme of this container for matches. A bird and rabbit hang on the sides with a hunter's horn in the middle. Incised on the back, "Sept. 27, 1870." Iron. American. Design patented by Frederick W. Brockspier. 5 1/2" x 9". $125-200.

Similar to the match container on the left. However, a stag's head and the figures of two hunters have been added to the match holder. The game, a rabbit and a bird, are hanging on the sides, and a hunter's horn is in the center. Iron. American. 5 1/2" x 10 3/4". $150-225.

A shuttered window is the design for this iron match container for the wall. The open top window box has the inscription "MATCHES" on it. Marked on the back "pat apld for." American (c. the last quarter of the nineteenth century). $100-150.

Nickel plated iron match container for the wall with a double open top pouch to hold the matches. The criss cross design on the front provides friction for striking the matches. A hunter's game, a rabbit and a bird, is hanging on the sides. A gun is slanted across the middle. 6 1/2" x 10 1/2". *Courtesy of Irontiques*. $195-225.

Left:
A shield, flags, drums, and a horn are part of the elaborate decoration of this wall match holder. The open topped drums are the containers, one for the good matches and the other for the burnt stubs. Iron. 5" x 10 1/2". $135-195.

A double match holder for the wall with one compartment over the other. The top bin is open for depositing burnt matches while the bottom container has a hinged lid to store the flammable matches. The striking surface is on the bottom and sides of the match container. Marked under the lid "Pat'd Jan. 10, 1872." Iron. American. 4 1/4" x 7". $150-225.
There is another version of this match holder (not shown). Instead of a floral design on the lid, it has the following advertisement: "J. H. HARTKE HARDWARE DEALER N. E. CORNER 8TH & MAIN STS. LOCKSMITHING BELL HANGING."

A single wall match holder with a hinged lid. A flower is on the six corners and a floral design covers the entire container. The striker is on the sides and underneath the match holder. Iron. American. 3 1/2" x 4 1/2". $75-120.

A two compartment match holder. The top receptacle for matches sits above the bottom receptacle which is used as a depository for the burnt stubs. The striking strip is between the two compartments. Nickel plated over brass. 3" x 5 1/8". $85-125.

A single compartment match holder for the wall. "MATCHES" is inscribed on the front. There is a corrugated strip across the lower front for striking a match. There are also striking surfaces on the sides of the receptacle. Iron. American. 4 x 6 1/4". $100-175.

Two urns on a shield form this match container for the wall. The urns are open at the top and have a wreath decoration on the front. The striking surface is on the lower face of the back plate. Marked on the back "Pat. appl'd for" and in an oval "Bradley & Hubbard West Meriden, Conn. 1800." Iron. American. Patented Jan. 28, 1868. 4" wide, 4 1/8" tall. $125-175.

A decorative match holder for the wall with a single open compartment. Shield shaped and covered with a scroll design. On the sides are small striking pads. American (c.1875). $75-125.

A beautifully detailed match container to hang on the wall. Acorns and oak leaves surround a hunter's pouch that holds the matches. Below is a boar's head and a pair of crossed guns. The hilt of a knife is on the left side and a hunter's horn is hanging at the bottom. The front of the bag has a rough surface for striking a match. Marked on the back "E. G. Zimmerman. Hanau, Germany." Iron (c.1865). 5 " x 6 3/4". $200-300.

A double match container with two open compartments, one for the matches and the other for the burnt stubs. The hanging hole is above a decorative cut out design. The striking panel is located between the match bins. Marked on the back "Pat'd May 10, 70." Iron. Made by Judd mfg. Co., New Haven, Connecticut. American. 3 1/2" x 7 1/2". $125-195.

Papier maché match container for the wall. Oriental characters appear in the background as well as on the front of the match holder. The striker is located on the lower front. Black, white, and touched with gold highlights. England (c.1880). *From the Kurschat collection*. $100-135.

A glass match container for the wall in the shape of an elephant's head. The opening at the top holds the matches. The striking surface is the diamond shaped area on the forehead of the elephant as well as the ribbed surface on the trunk. Made by King Glass Co., Pittsburgh, Pa. (c.1885). $75-125.

A single compartment match holder in the form of fireplace bellows. The ribbed surface on the front of the container is the striker. White metal. 1 7/8" x 5 1/2". $75-125.

A match container for the table or the wall.in the form of a stove. On the front is inscribed "COOK WITH GAS." There is a striker on each side. This is an unusual construction; the front of the stove is made of iron while the sides are tin. $100-125.

A single compartment match container attached to a shield shaped back plate. Embossed with an ornate pattern over all. The striking surface is located on the sides of the container. Iron. 4 1/2" x 5 5/8". $95-145.

A ceramic wall hanging match holder. "Matches" is written on the front. A ribbed rectangular striker is located below the match container. White with metallic gold trim. $35-55.

A masonic emblem is the theme of this match holder to hang on the wall. The "G" in the center is for geometry. The square represents tools of the trade. Motto on the side scrolls reads "Hope, Charity, Love, Truth." Iron. American. $125-175.

An oblong match container for the wall with the following advertisement: "TELEPHONE 32 UNION. FRANK HALL. HOMESTEAD N. J. COAL WOOD & FEED." The match bin is located at the top with a striker across the front. The three pegs on the bottom are used to hold keys. Iron. American. 4 x 8 3/4". *From the Sanders collection*. $150-195.

Right:
The open work at the top of this match container is reminiscent of a Gothic cathedral. Is the figure of Victor Hugo's Quasimodo, the Hunchback of Notre Dame? Another version was made without the spires at the top. Iron. American. 3 1/2" x 7 3/8". $150-200.

A beautifully beaded hanging match holder with two pockets. The colorful beads are stitched to a canvas background. "Matches" is spelled out above and the date "1904" is below. American. $75-110.

A stamped brass match container for the wall with the sides and bottom edges rolled back. The background is embossed with faces of cherubs. The match container displays a hunter and his dog on the front. This is rather an odd combination of themes. $45-65.

This match holder is bristling with armor, battle axes, spears, and a sword. A helmet with a bird tops off the background. The single match holder, in the shape of a shield, is open to hold the matches, and has a rough surface on the front to function as a striker. Marked on the back in an oval "Bradley & Hubbard West Meriden, Conn." American. 4 1/4" x 6 3/4". $150-225.

A plumed helmet forms this open top match container. The back plate is a highly decorated shield. A ribbed oval striking plate is at the bottom. Marked on the back "B 42." Iron with a copper finish. 4" x 6 3/4". $125-175.

A match container in the form of a bucket sitting on a shelf. The background is a brick wall. The under side of the shelf is divided into two parts; one side has the striker surface and the other has an etched decorative design. White metal. 2 1/2" x 4". $75-120.

This delightful match holder is unmarked as to origin. An Irishman wearing a hat is sitting above a basket smoking his pipe. The basket is the receptacle for the matches. The scrolled panels on each side have a ribbed surface to be used as strikers. White metal. 5" x 8 3/4". $150-225.

A lovely garden arch with a bird perched on each side forms this match holder for the wall. Ornate open work background with a musical instrument resting on one side. The urn in the center is open at the top to hold the matches. The striker panel is located across the lower front. Bronze. 4 3/4" x 8 1/4". $150-250.

A double match container for the wall with an open work design at the top and bottom. The back plate is made of iron while the open top match bins are made of tin. The striking plate is in the middle. American. 4 1/2" x 5 3/4". $65-95.

A wall hanging match container featuring a girl carrying two milk pails. The pails hold the matches. She is attired in typical Dutch dress, including a bonnet, apron, and wooden shoes. "TV" is embossed on the buckets. At the bottom is a bin to hold the burnt matches. Iron. 5 1/4" x 8". $125-150.

A single match holder for the wall with a hinged lid. Decorated with four diamonds on the container and beading around the edges. A ribbed striker is located across the lower front. The information on the patent application for this match holder states "In withdrawing the matches from the box, all that is required is simply to pass the thumb and finger through the opening and raise them out of the box, the lid being raised under the upward movement of the thumb and finger and closing by its own gravity, as the thumb and finger are drawn from the box." Marked "C. PARKER PAT OCT. 6, 1868." Iron. American. 4 3/4 x 5 1/2". $200-300.

A window with a wooden shutter is the theme of this match holder for the wall. "MATCHES" is written on the front of the open top container. A tree branch is in view in the top window pane. On the back is "101 Pat apl'd for." Iron. American. 4" x 5". *Courtesy of Irontiques*. $225-275.

A match container in the shape of a tiny basket to be hung on the wall. The rough surface of the basket can be used to strike the match. Brass. England (c. last quarter of the nineteenth century). 2" x 4". *From the Kurschat collection*. $125-175.

An iron frame for a mirror or picture forms this single open top match container. The striking strips appear on each side. Dated on the top edge "July 21, 1863. " "MATCHES" is embossed on the front of the receptacle. American. 4 3/8" x 7 1/4". *Courtesy of Irontiques*. $195-225.

A match holder for the wall or the table with two separate compartments with hinged lids. Cast iron body with tin lids. American (c.1875). 2 5/8" x 6 1/4", 3 1/2" tall. $75-95.

A pair of boots hanging on a rough board fence forms this match holder. A "FOR SALE" sign is tacked up below the boots. Heavy iron. *Courtesy of Irontiques*. $150-250.

A double match holder for the wall. The upper bin has a lid, and the one on the bottom is open for the burnt matches. Embossed on top is "Vestibule, Pat April 2, 89." A diagonal design cuts across the front and there are sprigs of shamrocks mixed with dots and crosses. The striking surfaces are located on the bottom of each receptacle. Iron. American. 3 1/4" x 6 3/4". *Courtesy of Judy Andrychuk*. $250-300.

A pair of boots hanging on a nail forms the design for this match holder. There is a mouse on top of the left boot and a mouse poking through a hole on the right boot. The back plate is covered with a rough diamond and dot design. The separate oval plate at the bottom is ribbed to form a match striker. Marked on the back "9'9." White metal. 3 3/4" x 7 1/4". $175-250.

A single compartment match holder to hang on the wall. A plain little container with a hinged lid. Above the box is written "PATENTED JULY 16, 78." The striking surface is on the bottom of the receptacle. Tin. American. 3" x 4 1/4". *From the Breon collection*. $45-65.

A single match holder for the wall advertising "E. C. SIMMONS KEEN KUTTER CUTLERY AND TOOLS." Iron. American. 4 1/4" x 6 3/4". *From the Sanders collection*. $125-175.

An advertisement covers this match holder for the wall. The message reads "TURN OFF THE GAS WHEN NOT USING, MATCHES ARE CHEAPER THAN GAS. PAT. JUNE 13-1899. COMPLIMENTS OF THE GAS CO." The striker is across the front of the bottom bin. Iron. American. 4 1/4" x 7". *From the Sanders collection*. $125-175.

An unusual match holder with a hook at the bottom to hold a pot holder. The small bin at the top holds the used matches while the large container below has a hinged lid for storing the unused matches. The message embossed on the top reads "MATCH BOX BRACKET, PATENTED, OCT 17, 1876." Iron. American. 3 3/4" x 7". *From the Breon collection*. $200-300.

A mountain lion is displayed below a Winchester rifle. The single open match container is decorated with stars and bars. The striking surface is underneath the match container. Iron. American. 3 3/4" x 4 1/8". *From the Gersten collection*. $95-125.

Lovely decorative open grill work background with a single open match holder. The match container is embellished with entwined branches and leaves with a diamond design around the edge. Iron. American. 5" x 7 5/8". *From the Sanders collection*. $125-175.

A single open match container to hang on the wall or stand on a table. Decorated with a large acorn. Marked on the bottom "NGL." Iron. American. 4 1/2" x 6 1/2". *From the Sanders collection*. $65-95.

A wall hanging match holder with two receptacles, one over the other. The upper is designed to hold a small match box and the lower to hold used matches. Tin with copper coloration. Made in England. 3 7/8" x 9". *From the Alperton collection*. $65-95.

A large oak leaf is the background for a single acorn that is the match holder. The rough surface of the leaf provides the striker. Iron. American. 4" x 7". *From the Sanders collection*. $75-125.

A match holder for the wall with two open receptacles. A primitive sculptured form of George Washington is featured on the back panel. Hand forged heavy iron. American. 4" x 9". *From the Alperton collection*. Extremely rare. Price unavailable.

A double open top match container with a striking bar across the front. One bin has "GOOD" on the front and the other, "BURNT." The word "MATCHES" was cut out of the background and a luminous piece of paper was affixed to the back, allowing one to easily locate the matches in the dark. Tin. American. 4 5/8" x 5". *From the Alperton collection*. $120-145.

A match container for the wall with two receptacles, one above the other. The larger one is for the matches and the smaller for the burnt stubs. The striking surface is at each end of the top holder. Albert D. Judd, of New Haven, Conn., patented this holder in 1870. His company's catalog of 1872 shows two versions of this piece. Iron. American. 3 1/4" x 6 5/8". *From the Alperton collection*. $95-150.

A match container for the wall with a single open receptacle. Embellished with open work leaf decorations. The striking surface is located on the underside of the holder. This match holder was featured in an 1884 advertisement. Iron. American. 5" x 7". *From the Alperton collection*. $125-175.

A match holder with two containers, one above the other. The upper one has a lid and the lower one is open and in the shape of a shell. Just above the shell is a shield shaped area with a message which reads "BURNT MATCHES 1873." The striking surface is under the top holder. Iron. American. 3 1/4" x 5 3/4". *From the Alperton collection*. $95-150.

A single match container with an open work back ground and the head of a Chinese nobleman in the center. The striking surfaces are located on both ends and underneath the match receptacle. Iron (c. last quarter of the nineteenth century). 3" x 4". *From the Alperton collection*. $75-100.

A match container for the wall with two open receptacles, one over the other. The strikers are on both ends of the upper container. This elaborate holder is completely embellished with floral, geometrical, and bird designs. Brass (c. late nineteenth century). 4 1/2" x 8 1/2". *From the Alperton collection*. $125-195.

A wall hanging match container to hold a small box of matches. An ornate pattern covers the entire surface. The sides of the match container are open to expose the striking surface on the box of matches. Iron. American (c.1900). 3 1/2" x 4 1/2". $55-80.

Right:
A single compartment match holder for the wall with a decorative pattern covering the entire surface. There are match strikers on the sides as well as the bottom of the container. Iron. American (c. the third quarter of the nineteenth century). 3 1/2" x 4 1/4". *From the Alperton collection*. $60-85.

A cast iron match holder for the wall. Rough boards form the background. There is a hook at the top to hold a key. The match container at the bottom is in the shape of a feeding trough. The ribbed strikers are on both ends. The number 2691 is impressed on the back. American. 3 1/2" x 7 3/4". $85-125.

A single compartment match safe. Embossed on the lid "C. PARKER PAT SEPT 14, 1860 & MAY 3, 1870." The ribbed strip across the lower front is the match striker. Iron. Patented by Hiram Richmond in 1869 and George Geer in 1870. 4" x 5 7/8". *From the Alperton collection*. $200-300.

This is an unusual match safe. It looks like a single compartment with just one lid, but opens to expose two bins. The back plate at the top reads "Pat. Nov. 2, 99." There is a striker on the bottom and top of the box. 3" x 3 1/4", 4" tall. *Courtesy of Irontiques*. $125-150.

A match container in the form of a coal bucket with an advertisement which reads "P. P. QUAYLE, 283 HUMBOLDT ST. & 116 CANAL ST. USE QUAYLE COAL & COKE." Iron. American. 5 1/4" x 3 7/8". *Courtesy of Irontiques*. $175-200.

A light weight brass match holder made to hang on the wall. The back plate has a crackled finish with five circles surrounding the single open match receptacle in the center. The striking plate is the ribbed strip under the container. American. 3 3/8" x 4 5/8". *Courtesy of Kossmann's Antiques*. $50-65.

A match holder in the form of a bust of Christopher Columbus. The open topped hat holds the matches. This match holder was most likely made for the World's Colombian Exposition that opened in 1893 in Chicago, Illinois and commemorated the 400th anniversary of the discovery of America. Iron. American. 2 5/8" x 4". $95-150.

A hammered brass match holder to be hung on the wall. The single open container on the front displays a scene of a hunter and his dog. 4" in diameter. *Courtesy of Irontiques*. $35-55.

Two upside down umbrellas are the containers for this match holder. One is for good matches and the other for used. The pebbled back plate is in the shape of a shield which is to be hung on the wall by the attached chain. There is an applied striking pad on the lower front. Brass. 6 3/4" x 7 3/4". $50-85.

A brick fireplace is the theme of this match holder for the wall. The half barrel on the hearth is the match container. The rough bricks are used to strike the match. Light weight brass. 3 1/4" x 4 3/4". *Courtesy of Irontiques*. $50-75.

A pair of boots with toes pointed toward a thermometer in the center is the design for this match container. The design at the top is a tree branch with a bird sitting on a nest of eggs. The background resembles cobblestones and is rolled at each corner. There is a small oval striker on the lower front. Copper. American. 7" x 7". *From the Kurschat collection*. $95-145.

A highly decorated double match safe. The top compartment has a hinged lid and on the front is embossed "MATCHES." The bottom container is open for the burnt stubs. Iron. American. Patented 1876. 4 1/8" x 7 3/4". *Courtesy of Irontiques*. $250-350.

A two tiered match holder with open top bins. A fanned spoke design and a star decorate the top and open lattice work appears on the back plate. There are striking surfaces on the sides and the bottom of the containers. American. 3 3/8" x 6". *Courtesy of Irontiques*. $200-300.

A cast iron double match holder. A lady with out stretched arms is holding her billowing dress which forms the two compartments of the match container. Marked on the back "Match Girl VIM." 4 1/8" x 6 1/2". $85-125.

A match container displaying a man's head. The open top of his hat holds the matches. Above the hat, a shovel, pitch fork, and a sickle are depicted. Bundles of grain are on either side of the man's head. The striker is on the lower front. Iron. American. Patented 1872. 4 1/2" x 6 1/4". $110-150.

A wall hanging match holder in the form of a coal bucket. The embossed advertisement on the back plate reads "JOHN M. REDDINGTON 99 WEST MAIN ST. TELEPHONE 390." The words "LEHIGH VALLEY COAL" appear on the bucket. There is a rough striking surface on either side of the bucket. Iron. American (c. the first quarter of the twentieth century). 4 1/2" x 6 1/4". *From the Alperton collection*. $175-200.

A wall hanging match holder with a single open receptacle. Ribbed striking strips are on both sides of the container. A decorative design covers the subject. Iron. American (c. the last quarter of the nineteenth century). 4 1/2" x 4 1/2". *From the Alperton collection*. $125-175.

A container to hold a small box of matches. The bottom receptacle is in the form of a shell and is used for the burnt matches. Ornate design in brass. England (c. early twentieth century). 3" x 8 1/2" *From the Alperton collection*. $125-175.

A truly unique match holder for the wall with an elaborate open work back plate. In the middle is the match container with a hinged lid. A long bin with a lid that has "BURNT MATCHES" embossed on the front appears to the left and a thermometer on the right. The match strikers are the ribbed areas on the front of the match holder and the bottom back plate. Iron. American. 5 1/4" x 6 1/8". *Courtesy of Irontiques*. $400-500.

A double match holder with an open design of flowers and leaves on the back plate. Two round containers with embossed leaves around the sides flank the oval striking pad in the middle. Iron. American. 6 1/2" x 9". $150-225.

Cast iron match safe for the table or wall. The hinged lid displays the embossed figure of a dog. Written on the front edge of the lid "PAT'D JAN. 21, 1862." Made by Mallory, Wheeler and Company, New Haven, Connecticut. American. 3 3/4" in length. *From the Alperton collection*. $125-150.

A laughing black boy wearing a large floppy hat is the subject of this match holder for the wall. The match containers to the left and right of the boy are ribbed on the sides to provide friction for striking the match. Iron. American. 4 1/4" in width, 4" tall. $95-125.

A flow blue match holder for the wall with a single compartment. This unusual match holder is designed to hang in a corner. Marked "Made in Germany." 5" tall. *From the Marsuobian collection*. $200-250.

This single compartment match safe with a hinged lid serves a dual purpose. It has four legs for standing on a table, and a hole at the top for hanging on the wall. The open work back plate and match container are decorated with scrolls and flowers. Iron. American. 3 5/8" in width, 3 3/8" tall. *From the Alperton collection*. $50-75.

A sheet brass match holder with curled corners. Embossed faces of cherubs cover the background. The design at the top is a bird sitting on a nest of eggs. A hunter and his dog are on the front of the match cups. American. *From the Leveen collection*. $50-75.

An iron match holder for the wall with an open match bin at the top. The decorative wreath has an open star shaped center. Three hooks at the bottom are for hanging keys. *From the Leveen collection*. $125-150.

A silver plated match holder for the wall. The head of a cat is the featured attraction. A round cup holds the matches and the tray underneath is a receiver for the burnt stubs. There is a ribbed striking surface on the lower front. Marked "Meriden B. Company. Quadruple Plate #50." Meriden, Connecticut. The back plate measures 2 1/2" x 8". *From the Aaronson collection*. $450-525.

A combination match holder and grooming tray. Match containers with a striking surface on the front are placed on either side of the mirror in the middle. The tray below is to hold a comb and brush. Tin plate. American (c.1900). 7 1/4" x 8". *From the Sims collection*. $45-65.

A combination match holder and grooming tray. A pair of open match holders are attached to the back plate and a mirror is placed in between. Embossed on the front of the large bin is "COMB & BRUSH." The front of the match holders are ribbed to provide friction for a match. Hammered tin plate. American (c.1900). 8 3/4" x 9". *From the Sims collection*. $45-65.

An aluminum match holder for the wall. The oval back plate is decorated with flowers and ribbons and has a scalloped edge. The square container holds a small box of matches and has a striker on the bottom. American. 3 5/8" x 5 3/8". *From the Sims collection*. $35-55.

A two compartment match holder for the wall. The back plate has a grained design and a decorative border with a beaded edge. The match container has an embossed ship on the front. Nickel plated sheet metal. 3 7/8" x 5 1/2". *From the Sims collection*. $45-75.

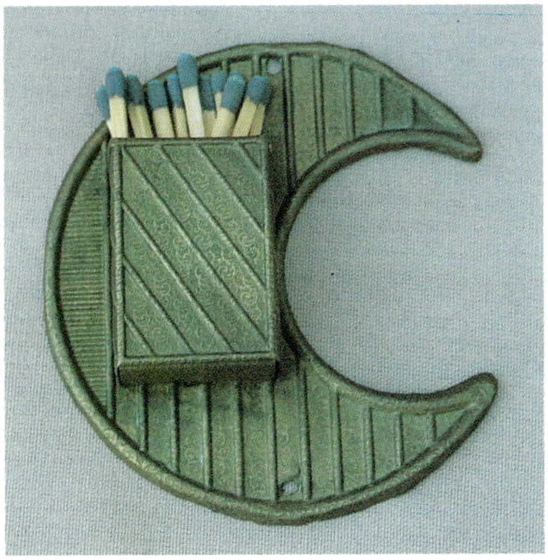

A single compartment match holder in the shape of a crescent moon, decorated with vertical and diagonal lines. The striker plate is located on the left side of the match box. Brass. 3 1/8" x 3 1/2". *From the Sims collection*. $35-65.

A shield shaped back plate holds two match containers in the shape of eggs with jagged tops. The embossed cobblestone design is the match striker. Brass. American. 4 1/8" x 4 1/2". *From the Sims collection*. $25-45.

This match container for the wall has a shelf to hold a pair of aluminum salt and pepper shakers. "MATCHES" is written above the open bin at the bottom. Tin painted blue. American. 5" x 6 7/8". *From the Sims collection*. $45-75.

A long tin match holder for the wall. The open bin at the bottom is marked "MATCHES," and above the top bin is written "BURNT MATCHES." The striking strip is at the top between the hanging holes. There is a message on the back that reads "PERFECT MATCH BOX Pat. Dec.1911." American. 7" tall. *From the Sims collection*. $45-65.

This match pocket for the wall holds 11" long fireplace matches. A tavern scene is displayed on the front. Inscribed on the back "Made In England." Brass (c.1965). 7 7/8" in diameter. $45-65.

A walnut match holder for the wall. Crossed boards with a cut out square at the top form the background for the single basket container. American (c.1875). 2 3/4" x 4". *From the Kurschat collection*. $55-75.

A match dispensing machine. The front of the machine is embossed "MATCHES 1 CENT." When a penny is inserted in the slot on the left side, a match will drop into the bin on the right side. On the lower front is a cigar cutter. Iron. 6" x 7 1/2", 13" tall. *Courtesy of Eve Naslund*. $700-800.

A carved eagle with a single match container. The match strikers were sandy surfaces glued to the wings (worn away in this photograph). The number 226 is engraved on the back. Walnut. American (c.1885). 4 3/4" x 7 3/8". *From the Kurschat collection*. $125-150.

A wooden match holder for the wall with a cut out design at the top. The single barrel shaped container is decorated with a dropped ornament at the bottom. Walnut. American (c.1875). 2 1/2 x 6". *From the Kurschat collection*. $65-95.

An ornate walnut match holder for the wall. The background is formed with loops of rough branches and ivy leaves in an open design. The goblet holds the matches. American (c.1875). 5 1/8" x 9 3/4". *From the Kurschat collection*. $80-135.

A shield with a dropped leaf design is the back plate for this match holder for the wall. There are two half barrels to hold the matches, one for good matches and the other for the burnt. The small shield in the center had a coat of sand glued to the front to act as a striker (missing in this photo). On the back is #37. Walnut. American (c.1885). 5 1/2" x 9 1/8". *From the Kurschat collection*. $95-125.

Sculptured leaves form the background of this combination match holder and shelf. Two urns with pointed drops hold the matches. The round shelf holds a candle stick. Walnut. American (c.1880). 8 1/4" x 12". The shelf is 6" in diameter. *From the Kurschat collection*. $125-150.

A wall hanging match holder in the form of a shrine. A single urn with a dropped acorn holds the matches. Written on the back "VARIETY BRACKET WORKS SOUTH BEND IND. 71." Walnut. American (c.1880). 4 3/8" x 10 5/8". *From the Kurschat collection*. $75-95.

An Eastlake style double match holder. The half barrels are the containers, one for good matches and the other for the burnt stubs. On the back is #264. Walnut. American (c.1890). 6 1/8" x 11 7/8". *From the Kurschat collection*. $85-120.

A single match holder is mounted on a bed of oak leaves. The barrel shaped match container is decorated with a dropped acorn. On the back is #22. Walnut. American (c.1875). 5" x 8". *From the Kurschat collection*. $65-85.

A shield shaped plaque with a decorative cut out design is the background for this match holder. Two half barrels with dropped acorns hold the matches. On the back is #37. Walnut. American (c.1875). 5 1/2" x 9 1/8". *From the Kurschat collection*. $95-125.

A wooden match container for the wall with an open work background with branches and ivy leaves. A pair of receptacles with pointed drops hold the matches. The striking surface is in the center. Walnut. American (c.1875). 6 3/4" x 11". *From the Kurschat collection*. $95-125.

A wooden match holder for the wall. A leaf design is carved on a solid background with a burning match carved in the center. A double match container with a criss cross design on the front. 4 3/4" x 7 1/4". *From the Sims collection*. $45-65.

A carved alder wood match holder for the wall. A round plaque with the head of a Chamois is displayed in the center and encircled by a French horn. The match holder is in the shape of a bird's nest and is adorned with acorns and leaves. On the back "A & UT" (c.1880s). 3 3/4" x 7 1/2". *From the Crabb/Nomann collection*. $325-425.

A wooden match holder for the wall. The advertisement reads "ALFRED H. SPOONER You can use a barrel of tacks, but will not need a barrel of money, if you trade with me. SHERMAN MILLS, MAINE." The barrel holds the matches. Across the bottom edge is a piece of sandpaper to be used as the match striker. American. 2 1/2" x 6". *From the Breon collection*. $60-95.

A wooden match holder for the wall with the message "With the compliments of FOARD & STOKES CO. Astoria, Oregon. Dealers in GROCERIES, HARDWARE, CROCKERY. Fresh Fruits and Vegetables. Ship Chandlery. WE BUY AND SELL EVERYTHING." American. 2 3/4" x 9 5/8. *From the Breon collection*. $60-95.

An Eastlake style match holder for the wall. Walnut accented with carved leaf decorations which are painted black. American. (c.1880). 7 1/2" x 11". *From the Crabb/Nomann collection*. $200-300.

A souvenir match holder from "Hot Springs New Mex." The picture at the top displays a windmill, and "Matches" is inscribed on the front of the box. American. 3 1/2" x 7 1/8". *From the Sims collection*. $25-45.

A single match receptacle for the wall. Open work background with a striking bar on each side of the urn. Advertisement reads "WILSON BRO. GRINDING MILL EASTON, PA." Iron. American. 4 1/4" x 6 1/4". *Courtesy of Irontiques*. $150-195.

A double match holder for the wall. The bottom compartment has a hinged lid for the unused matches and the top bin is open for the burnt stubs. Iron. American. Patented May 14, 1870. *Courtesy of Irontiques*. $195-250.

Pyrography, the art of burning a design on wood, was popular with the hobbyist in the late 1800s and early 1900s. This amusing match holder has a verse on the front which reads "Don't scratch matches on the walls scratch them on my overalls." The pants have a sandy surface for this purpose. The single cup on the right is the match container. Bass wood. American. 7" x 9". *From the Marsuobian collection*. $35-50.

Round match container with a single open compartment. Advertisement above the bin reads "BUY THE," and on the receptacle is "CAPITAL SQUARE STOVE." Around the outside edge is "D. H. PLIENH COOPERSBURG, PA." Iron. American. *Courtesy of Irontiques*. $150-195.

A lucky horse shoe over a star is the theme for this match holder for the wall. The barrel match container has the inscription "KINDLING WOOD" embossed on it. On the bar at the top is the date 1883. On the left side of the horse shoe is "Gut" and on the right side is "Guluick." On the left tip of the horse shoe is a "G" and on the right an "L." Iron. American. 4 1/4" x 5". *Courtesy of Ken and Barb's Antiques*. $125-150.

This cute match holder for the wall has many functions. "THE HANDY ELEPHANT IS USEFUL - DECORATIVE - CONVENIENT." Made to accommodate a pair of aluminum salt and pepper shakers, matches, and a hot pad that hangs on the elephant's back toe. Iron. Designed and manufactured by A. W. Reiser, Toledo, Ohio. American (c.1930s). 7" x 10". *Courtesy of Irontiques*. $225-250.

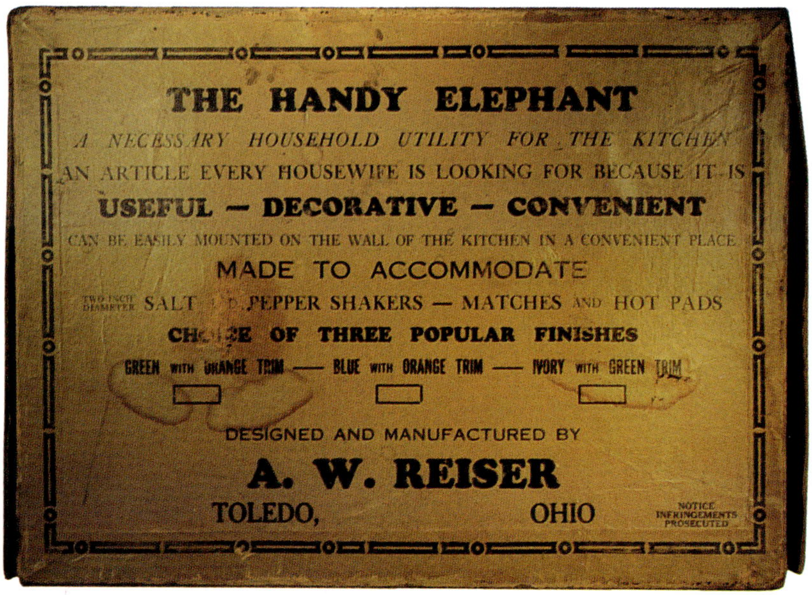

Box in which the elephant match holder was packaged.

A double match holder for the wall. Open grill work background with a large open container on the top and a small one on the bottom. Ovals on the front were usually painted with a floral design. There are striking surfaces on the bottom of the match bins. Iron. American. 3 3/8" x 7". *From the Gersten collection*. $95-150.

Hard to find match striker for the wall. Not very many were made without a match container. Brass. England. 3 1/8" x 5 7/8". *From the Gersten collection*. $35-65.

A lovely lacy double match holder for the wall. The open top bins are covered in an ornate design. The striking surface is on the ends of the match containers. Iron. American (c.1880). 4 1/4" x 7". *Courtesy of Irontiques*. $110-$150.

Acorns and oak leaves are the theme for this attractive match holder for the wall. A pair of open top acorns are the containers for the matches. The striking surface is between the two match bins and in the shape of an acorn. Small acorns adorn the top and bottom of the design. Iron. American. 5 3/4" x 8 1/2". *From the Gersten collection*. $150-225.

A double match holder for the wall with a lacy back plate. The two open bins are decorated with a sunburst design. Iron. American. 3 3/4" x 6 1/8". *Courtesy of Irontiques*. $110-145.

A double match holder for the wall. The top box has a lid and the bottom box is open. An ornate design with birds and tree branches covers the entire subject. Iron. American. 5 1/8" x 8 3/4". *From the Gersten collection*. $125-175.

A double match holder for the wall. The top bin is open for the burnt matches. The bottom compartment has a lid and holds the good matches. Embossed on the lid is "C. PARKER PAT SEPT 14, 1869 & MAY 3, 1870." Iron. American. 4 1/8" x 6 5/8". *Courtesy of Irontiques*. $250-300.

A boot on an artist's palette is the subject of this match holder for the wall. The palette is covered with embossed sea shells. The match striker is the ribbed oval below the boot. Brass. 3 3/4" x 5 1/4". *From the Kurschat collection*. $65-85.

A brass match holder to be hung on the wall. A double line frames a burning match that is engraved on the front of the match bin. 3 1/2" x 4 1/2". *From the Gersten collection*. $50-75.

A highly decorated and very pretty match holder for the wall. A brick pitched roof and chimney form the house. The figure of Punch with a large cigar in his mouth is leaning out of the window. The chimney is open to hold the matches. A fancy scroll around the sides and bottom sets off the design. England. 3 1/2" x 7 1/4". *From the Gersten collection*. $300-500.

A rather plain match holder for the wall. The match bin has a divider inside which forms two separate compartments, one for unused matches and the other for the burnt stubs. The ribbed striker runs down the middle. England. 3 1/8" x 5 7/8". *From the Gersten collection*. $75-95.

A double open match holder for the wall with a large match bin at the top and a small one on the bottom. Open work back plate with a clover leaf hanging hole. An ornate etched design covers the entire subject. The strikers are located on the sides of the match bins. Iron. American. 3" x 6 3/8". *From the Gersten collection*. $85-145.

An unusual bowed front match holder for the wall. The single match container is a half cup with a striking strip down the front. There are two hanging holes. Ornately decorated with a design of scrolls and flowers. Iron with a bronze patina. 4 3 /4" x 5 1/2". *From the Gersten collection*. $125-175.

This pretty single match holder has the figure of an angel embossed on the bowed front and an open work background with graceful scrolls. #319 is inscribed on the back. Iron with a bronze finish. 4 1/8" x 5 1/2". *From the Gersten collection*. $150-225.

A match holder for the wall with twin match bins. Covered with an attractive scroll and leaf design. The striking surface is located on the sides of the containers. Iron. American. 4" x 4 1/8". *From the Gersten collection*. $125-175.

A match holder with three open bins. Two are located above the oblong match striker. The container for the burnt matches is below. A wreath at the top encircles the portrait of a woman. Copper. 3 1/4" x 3 1/2". *From the Gersten collection*. $250-300.

A top hat attached to a horse shoe forms this match container. The advertisement reads "SHONFIELD & FRIEDRICH, ONE PRICE, CLOTHIERS." The bar across the middle reads "MANSFIELD, O." The tips of the horse shoe are rough to provide friction for the match. Iron. American. 3 1/2" x 4". $125-195.

The King of Clubs is the feature of this card shaped match holder for the wall. A ribbed striking surface is on both ends of the container. Brightly colored ceramic. 2 1/2" x 3 1/2". *Courtesy of "Before My Time" Antiques*. $60-85.

A simple open match container with a divider in the middle. One side is for the new matches and the other is for the burnt stubs. The striking surface is underneath the bin. Iron. American. *From the Alperton collection*. $45-65.

A bald eagle with spread wings is the feature of this beautiful match holder for the wall. A shield with stars and stripes surrounded by ribbons is the match container. Cast brass with a pewter shield. American. 5 1/8" x 8". *Courtesy of Ken and Barb's Antiques*. $195-250.

"TAM O' SHANTER" is engraved on the front of this match holder. A man is leaning over a barrel that holds the matches. Terracotta. Made in Austria. 4 1/2" x 6 3/8". *Courtesy of "Before My Time" Antiques*. $175-225.

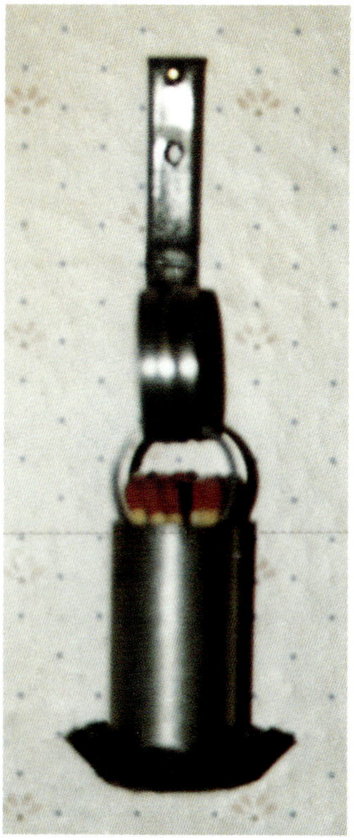

Far left:
A wall bracket with a ring to hold the match cylinder. The striker is on the bottom of the container. Stamped tin. 1 1/2 " x 5". *Courtesy and Ken of Barb's Antiques*. $18-35.

Left:
A single bin match holder covered in morning glories. There are strikers on the sides of the container. Heavy cast iron. 4" x 7". *Courtesy and Ken of Barb's Antiques*. $45-65.

A single compartment match holder with an open work back plate. There are strikers located on the sides of the match bin. Cast brass back plate with a cast iron match bin. American. 4 " x 4 1/2". *Courtesy and Ken of Barb's Antiques*. $45-75.

Embossed acorns and leaves cover this single compartment match holder. The back plate has three holes cut out in the shape of acorns. A striker is on the sides of the match bin. Cast iron. American. Patented 1876. 3 1/4" x 5 1/4". *Courtesy of Ken and Barb's Antiques*. $75-125.

A single cup match holder on an ornate back plate. The striker is located under the cup. Brass. 4 1/2" x 5 1/8". *Courtesy of Ken and Barb's Antiques*. $45-75.

A single container with an open top match bin. The advertisement reads "J L. HUDSON CLOTHIER Detroit Opera House Building." The striker is across the front of the match holder. Iron. American. 3 3/4" x 4 1/2". *Courtesy of Ken and Barb's Antiques*. $75-125.

Glass head of Lady Liberty with an open top for the matches. The striker is the scored section on the bottom. 2 5/8" x 4 5/8". *Courtesy of Ken and Barb's Antiques*. $75-95.

A decorative French match holder. A woman on one side and a man on the other, each hold a basket. The baskets act as the match containers. The striker is on the bottom. White bisque. 4 1/2" x 7". *Courtesy of Judy Andrychuk*. $75-125.

A double cylinder match container. The hanging hole is in the shape of a flower. Open work back plate with a fleur de lis on the bottom. Brass. 2 1/2" x 6 1/2". *Courtesy of Judy Andrychuk*. $35-65.

A basket divided into two compartments is the match container. The open work back plate is intertwined with branches and leaves. On the back "Patented 1868." Iron. American. 5 3/4" x 6 1/2". *Courtesy of Judy Andrychuk*. $100-175.

A lovely ornate double match holder. The bottom container has a lid and the top container is open for the burnt matches. The match striker is in the mid section of the back plate. Marked on the back "Adams and Westlake. Chicago & Philadelphia USA." Cast metal. American. 3 3/4" x 4 1/4". *Courtesy of Judy Andrychuk*. $95.150.

A double match holder with open scalloped edge bins. Decorated with an impressed design on the back plate. The striker plate is across the bottom. Has a top and bottom hole for securing to the wall. Steel with remnants of brass plating. 3 1/4" x 5 1/2". *Courtesy of Judy Andrychuk*. $65-85.

Far left:
A lady in a long blue gown wearing a necklace is the back plate for this single match container. "MATCHES" is embossed on the front of the open compartment. The striker is across the bottom of the holder. Tin. 3 1/8" x 5 1/4". *Courtesy of Judy Andrychuk*. $50-75.

Left:
A rectangular holder for a box of matches. The sides have open slots so the match striker on the box is exposed. The woven design on the front may also act as a match striker. On the back "T4198-No.2-S." Brass. 2 1/2" x 4 1/4". *Courtesy of Judy Andrychuk*. $35-50.

An ornate match container in an intricate open work design with two open match bins, one above the other. The striker is across the front of the lower bin. Brass. 3 1/2" x 6". *Courtesy of Judy Andrychuk*. $80-120.

A hand carved open match pocket for the wall with a criss cross pattern. The hunter's game, a goose, is hanging on the side. A branch with leaves adds to the decoration. Walnut painted gold. 5" x 9". *Courtesy of Judy Andrychuk*. $50-95.

A match holder with a thistle in a circle on the front. "Matches" is written across the top. The striker is the ribbed rectangle at the bottom. Brass. 2 1/2" x 5". *Courtesy of Judy Andrychuk.* $65-95.

A sailor dressed for stormy weather is at the top of this match holder. A coastal scene with a windmill and a cloudy sky covers the lidded match bin. The striker is the oblong section on the lower front. Tin. 4 1/2" x 4 1/2". *Courtesy of Judy Andrychuk.* $50-85.

A beautifully detailed ornate match container for the wall. An eagle with spread wings is perched on top of a quiver of arrows. A trumpet flower is in the middle with a dog on each side. The basket below is the match container. The rough surface of the basket is the match striker. Cast iron. 5 3/4" x 9". *Courtesy of Judy Andrychuk.* $150-225.

Grapes and leaves cover the entire surface of this match holder. Open work back plate with grapevines entwined with leaves. Match strikers are located on the sides of the container. Cast iron painted gold. 3 1/2" x 5 1/2". *Courtesy of Judy Andrychuk.* $95-145.

A single match compartment in the form of an acorn against an oak leaf. Aluminum. 3 1/2" x 6 1/2". *Courtesy of Judy Andrychuk*. $25-45.

This attractive contemporary match holder with houses, trees, and a large sunflower on the bottom is for long fireplace matches. Brass. 3" x 9 3/4". *Courtesy of Judy Andrychuk*. $30-45.

Advertising tin for "SCHULZE'S BUTTER-NUT The Bread Of Quality" with a single open match container. The striker is across the bottom of the bin. Tin. American. 3 1/2" x 4 3/4". *Courtesy of Judy Andrychuk*. $45-75.

Advertising tin for "MILWAUKEE HARVESTER CO. Binders and Mowers Always Reliable." Single open match container with the striker underneath. Tin. American. 3 1/2" x 4 3/4". *Courtesy of Judy Andrychuk*. $45-75.

Advertising match holder with a single bin. Advertisement reads "Compliments of JETTINGHOFF & BECKMAN Delphos, O. Phone 64. Complete House Furnishers." The striker is on the bottom of the container. Tin. American. 3 1/4" x 5 1/2". *Courtesy of Ken and Barb's Antiques*. $45-75.

Advertising match holder with a single open cup. Message reads "DRY GOODS, CLOTHING, SHOES AND GROCERIES. St. Charles Phone 2023." Striker is under the cup. Aluminum. American. 2 1/4" x 3 1/2". *Courtesy of Ken and Barb's Antiques*. $25-35.

Advertising match holder with a single bin. A four leaf clover is in a circle with W.S.S.C. on the leaves. Message reads "Clover Brand Shoes For Wear-Style-Service-Comfort." The striker is underneath the match bin. Tin. American. 3 1/2" x 4 3/4". *Courtesy of Ken of Barb's Antiques*. $45-75.

This unusual advertising match holder has the photos of the proprietors in an oval frame with their names underneath. The back plate is in the form of a picture frame with flowers in each corner. The single match bin has the message "J. G. Jones & Co. Match All Competition And Are Scratching For Your Trade Kenes, Nebraska" written on the front. "Hardware, Shoes And General Merchandise" is written on the sides. Aluminum. American. 3 3/4" x 6". *Courtesy of Ken and Barb's Antiques*. $75-125.

Holders For Large Boxes Of Matches

In the early 1900s, matches were sold in a cardboard box with a striking surface on the side. These boxes held about 200-250 large wooden matches. As a result, a different kind of match holder was needed, one that would accomodate for these larger boxes of matches. These containers usually have a slot cut out of the side to expose the striking surface of the match box. They became very popular and could be found in nearly every kitchen in America. Today, box matches are still in production and a few wall containers are still being made to hold them.

A tin container for a large box of matches. Advertisement on the front reads "You Can't Match Us For Electrical Service CHRISTENSON RADIO & ELECTRIC." On the back is written "Made by Newton Manufacturing Co. Newton, Iowa." American. 2 3/4" x 4 1/8". *From the Sims collection.* $45-65.

A match holder for the wall made to hold a large box of matches. The logo on the front of the match container shows a hand striking a match on the side of a match box. The message reads "BIRD'S EYE MATCH BOX HOLDER For Non Poisonous Bird's Eye Matches." Tin. American (c. the first quarter of the twentieth century). 4" x 6 1/4". *From the Sims collection.* $95-125.

A tin match holder for a large box of matches. Painted white with red strawberries. This type of match container was inexpensive and came in many colors, plain or with various designs. American. 3 1/4" x 5". *Courtesy of Irontiques.* $20-30.

A tin match holder for the wall. The hinged lid opens to reveal two compartments. The right side has a section to hold a large box of matches. On the left side of the lid is a slot to drop the used matches into the bin below. The striker is on the lower left-hand side. Painted pink with a rooster and flowers as decoration. American (c.1935). 4 1/4" x 5 1/2". *From the Sims collection.* $45-75.

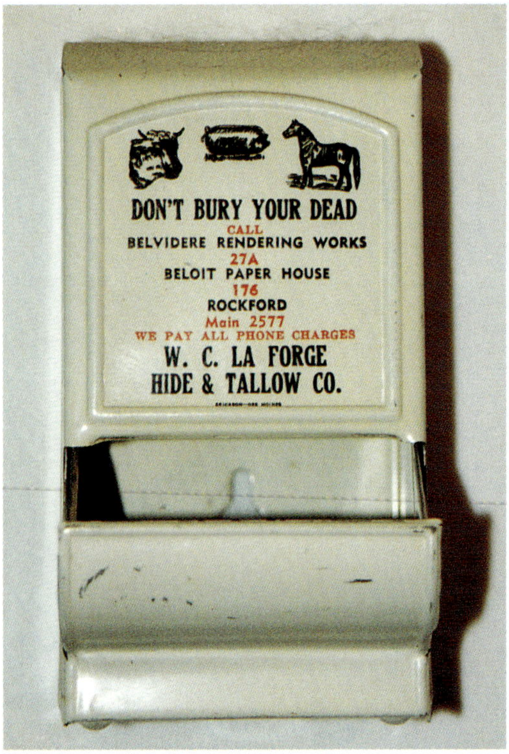

This match container holds large kitchen matches. The advertisement reads "DON'T BURY YOUR DEAD Call Belvidere Rendering Works 27A, Beloit Paper House 176, Rockford Main 2577, We Pay All Phone Charges, W. C. LA FORGE HIDE & TALLOW CO." Tin painted white. American. 3 1/8" x 6". *Courtesy of Ken and Barb's Antiques.* $35-55.

"THE ECONOMY SELF FEEDING MATCH BOX HOLDER, Patented June 29, 97." by Economy Match Box Co. Chicago.The instructions on how to use the holder are printed on the front: "Place extended shank of holder between bottom of tray and sliding cover of match box. Push tray to bottom of holder. Then hang up. The matches will feed out until box is emptied." Tin. American. 2 3/4" x 6 1/2". *Courtesy of Ken and Barb's Antiques.* $35-45.

A match holder to hold large kitchen matches with advertisement for "Dr. Pepper Drink a bite to eat at 10-2-4 o'clock. Keep Dr. Pepper In Your Home." Tin painted green. Made by P. & N. Co., Fulton, Illinois (The stamping dyes were destroyed in 1942, so this item can no longer be made). American. 3" x 6". *Courtesy of Ken and Barb's Antiques.* $75-95.

A ceramic match box holder colorfully decorated with cross stitching. The top half is covered with birds and flowers. "MATCHES" is printed on the front. The top bin is used for the burnt matches. The rough embossed section on the lower front is the match striker. $60-95.

An old unpainted tin holder for large kitchen matches with an open top. The compartment at the bottom has a hinged lid. "MATCHES" is embossed on the front. American (c.1910). 3 7/8" x 4 5/8". *From the Sims collection*. $35-65.

This match holder for large kitchen matches has an open bin at the bottom. The advertisement reads "A Little 'Thank You' From RIGGINS MERCANTILE General Merchandise 'The Biggest Little' Store On Salmon River RIGGINS - IDAHO." Tin painted white with red decorations. American. 2 3/4" x 5 1/8". *From the Sims collection*. $35-65.

This match container for the wall has a closed top and an open lower bin. On the front is a basket of fruit and vegetables and an advertisement that reads "Compliments of R. J. WEISER CASH STORE 'A Square Deal for All' Odin - Kans. Phone 18F23." Tin. American. 2 3/4" x 5 1/8". *From the Sims collection*. $35-65.

A match holder for the wall. Made to hold a large box of matches. The tray on the left side is for the burnt stubs. Embossed on the top "KREAMER." The decal showing a man wearing a sombrero and serape and pulling a stubborn mule was probably added at a later date. Tin. American. 5 1/8" x 6 7/8". *From the Sims collection*. $35-65.

This colorful series of match box holders is contemporary, but charming and collectible. Most are marked "Jasco Hong Kong."

Left:
A dock side scene with a beached boat is the topic of this match box holder for the wall. There is an opening on the sides to expose the striking surface on the match box. Marked on the back "Jasco Hong Kong." Tin. 3" x 6 1/4". *From the Sims collection*. $25-45.

Right:
A rooster with hens in front of a chicken coop is the theme of this match holder. Chicken wire frames the opening on the side which is cut out to accommodate the striker on the match box. The top and bottom bins have hinged lids and are painted green with gold trim. Not marked, but part of the Jasco series. Tin. 3 1/8" x 6 1/8". *From the Sims collection*. $25-45.

Left:
A scenic match box holder for the wall, featuring a doe with her fawn at a watering hole. Pine trees in the background and birch trees on the sides give the impression of a forested glen. Not marked, but no doubt one of the Jasco Hong Kong series. Tin. 3" x 6 1/4". *From the Sims collection*. $25-45.

Right:
A pair of ducks taking flight against a background of reeds and cattails sets the scene for this match holder for the wall. The hinged lids are trimmed with gold. There are openings on the sides to expose the striking surface on the match box. One in the Jasco series. Tin. 3" x 6 1/2". *From the Sims collection*. $25-45.

Left:
A long eared dog sitting under a tree holding a butterfly net is the subject of this match box holder. There are slots on the sides to expose the striking surface on the side of the match box. Hinged lids with a gold trim. Marked on the back "Jasco Hong Kong." 3 x 6 1/2". *From the Sims collection*. $25-45.

Right:
A ceramic match box holder decorated with cream glaze and a design of green grapes and grape leaves. Open at the top to insert a box of matches. Japan. 3 3/4" x 5 3/4". *From the Sims collections*. $25-45.

Left:
"Hickory Dickory Dock, Three Blind Mice" and "The Cat And The Mice" are titles of books on the shelf on the sides of this match box holder. On the front, there is a mouse at the top of the clock and another is setting the time. Hinged lids with gold trim. There are openings on the sides to expose the striking surface on the match box. Marked on the back "Jasco Hong Kong." Tin. 3 1/8" x 6 1/8". *From the Sims collection*. $25-45.

Right:
A ceramic match box holder for the wall with a strutting rooster on the front. Written on the back "Poppy Trail By Metlox Made in California." White glaze with a yellow, red, and green rooster and a green knob on the drawer. Open at the top and bottom. American. 4 1/4" x 7 3/4". *From the Sims collection*. $75-100.

A hammered copper match box holder. Two pine cones hanging from a twig decorate the front panel. On the back is "Gregorian Made In The U.S.A." and a paper label that reads "Solid Copper Made In U.S.A." 3" x 7 1/2". *From the Sims collection*. $25-45.

This container to hold a box of matches is open at the top and bottom. On the front is embossed "MATCHES" and the logo "COCA-COLA Delicious And Refreshing." Tin. American. 3 1/4" x 6 1/4". *From the Sims collection*. $35-65.

A copper match box holder for the wall with a pierced design on the front. The slots on each side expose the striking surface on the box of matches. American. 3 1/4" x 10". *From the Sims collection*. $25-45.

A cabinet filled with kitchen items is the subject of this match box holder. Colorfully painted on a brown wood grain background. On the back is "Hallmark Cards Inc. 300EBC68-8 K. C. Mo. 64141 Made In The U.S.A." Tin. American (c. early 1970s). 3 1/8" x 6". *From the Sims collection*. $35-45.

A yellow ceramic container for the wall which holds a large box of matches. A chef is sitting above an open match bin. 3 1/2" x 5 3/8". *From the Sims collection*. $25-45.

This wood match container for the wall holds a large box of matches. A colorful painted rooster is framed with a design of scallops and dots. Made in Japan. 3 1/2" x 7 1/8". *From the Sims collection*. $20-35.

A wooden container for the wall which holds a large box of matches. A colorful stylized rooster is standing between two sprigs of red daises with leaves. Single open bin with a notch on the front to facilitate the removal of the matches. 3 1/8" x 6". *From the Sims collection*. $20-35.

Tin Advertising Match Holders

Beautifully lithographed tin match holders were a perfect place to advertise a product. They were inexpensive to manufacture, costing just pennies, so most were gifts to customers. A match holder hanging at eye level used on a daily basis was an excellent way to promote a product. The ads for stoves, flour, coffee, and cream separators, for example, appealed to the housewife. For the men, farm machinery, whiskey, and cigars were often displayed. Medical problems were attacked with grandiose promises of instant cures for any and every ailment. A few tin match holders with advertisements were made in the late 1800s, but the majority were made in the early 1900s.

This advertising match holder was designed for Old Judson whiskey. The scene depicts a father coming home to a warm fire. His dutiful wife is removing his coat and his sweet daughter in a blue dress with a bow in her hair is handing daddy a glass of — whiskey! The copy above the illustration reads "Old Judson." On the match container "J. C. STEVENS, 518 DELAWARE ST., KANSAS CITY, MO. U.S.A." The striker is across the bottom. Beautifully lithographed tin. American. 3 1/2" x 5". *From the Alperton collection*. $175-250.

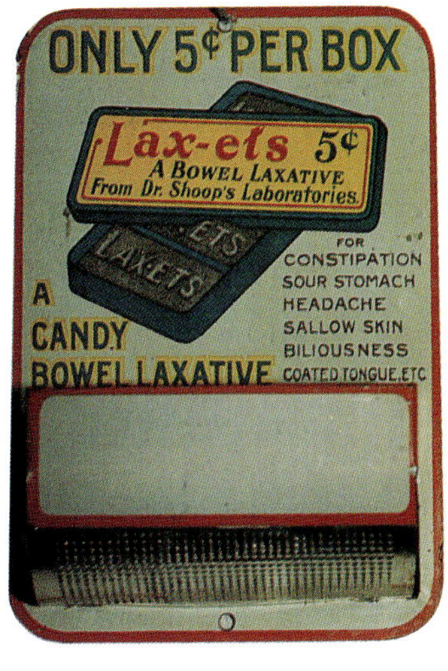

This tin match holder for the wall promises a great deal. The advertisement reads "Lax-ets A Bowel Laxative from Dr. Shoop's Laboratories For constipation, sour stomach, headache, sallow skin, billiousness, coated tongue etc." On the side of the container "Dr. Shoop + Racine, Wis." The striker panel is on the bottom of the box. American. $75-100.

A match holder designed for the Wrigley Company. The back panel shows an illustration of Mr. Wrigley, and the copy reads "THE MAN JUICY FRUIT MADE FAMOUS." Written in small type "Wrigley — Chicago." The striker is on the lower part of the receptacle. Tin. 3 1/2" x 5". *From the Alperton collection*. $125-195.

A match holder advertising "CERESOTA Prize Bread FLOUR Of The World." In the circle is written "The Northwestern Consolidated Milling Co. Flour." The boy on the top is slicing a large loaf of bread. The match container is the square box with striking surfaces on the sides. American. 2 1/4" x 5 3/8". *From the Breon collection*. $200-500.

Another version of a match holder with advertising for the Ceresota Flour Co. The barrel functions as the match holder and there is a piece of sandpaper on the bottom for a striker. A little boy is sitting in a chair slicing a large loaf of bread. Beautifully lithographed tin. American. 2 1/4" x 5 3/8". *From the Breon collection*. $200-400.

This match holder depicts a dairy farm and a woman using a "Tubular CREAM SEPARATORS." The text reads "The Sharples Separator Co., West Chester, Pa. - Toronto, Can.- Chicago, Ill." The striker is on the lower front. Tin. American. 2 1/8" x 6 1/2". *From the Alperton collection*. $200-350.

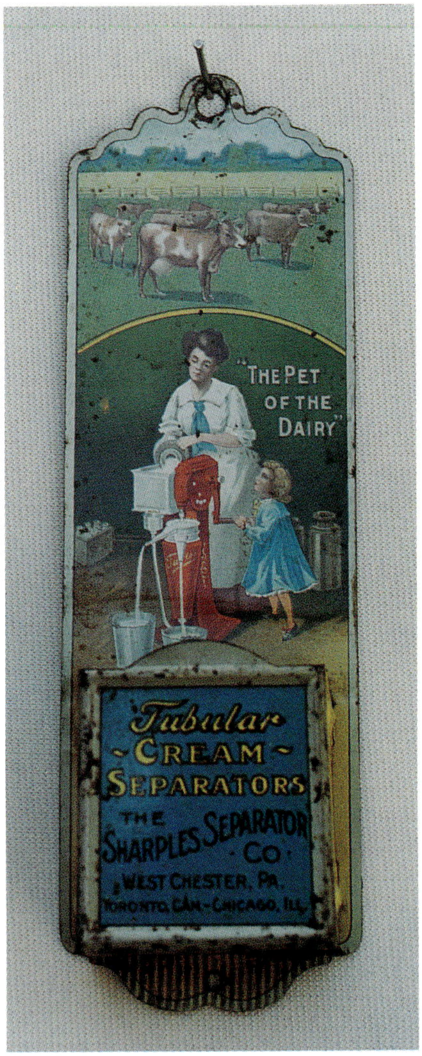

Beautifully colored tin match container featuring a dairy farm and a woman using a cream separator. In this newer version of the "Tubular CREAM SEPARATORS," a little girl and "The Pet of the Dairy" is added to the design. The text on the receptacle reads "THE SHARPLES SEPARATOR CO., WEST CHESTER, PA. TORONTO, CAN.-CHICAGO, ILL." The striker is located on the lower front. Tin. American. 2 1/2" x 6 3/4". *From the Breon collection*. $200-300.

This match holder for the wall was designed for the DeLaval Separator Company with two receptacles, one for unused matches and the other for used. The company kept a running count of the number of separators sold. This one reads "1,000,000 in use." Painted black with blue trim. Tin. American. 4" x 6 1/4". *From the Alperton collection*. $200-350.

This cardboard match holder advertises Garland Stoves and Ranges with a calendar attached dated June 1904. The text informs us that "IF ITS A 'GARLAND,' THAT'S ALL YOU NEED TO KNOW ABOUT A STOVE OR RANGE" and "'GARLANDS' WARM THE HOMES, BAKE THE BREAD AND ROAST THE MEATS THAT MAKE THE MAN." There is a strip of emery paper on the front of the match container to be used to strike the match. American. 3 3/4" x 6 1/2". *From the Breon collection*. $100-150.

Another version of a match holder designed for the Garland Stoves and Ranges Company. On the upper part of the back panel "OFTEN IMITATED - NEVER EQUALED. THE WORLD'S BEST." On the front of the receptacle "THE MICHIGAN STOVE COMPANY Largest Makers of Stoves and Ranges in the World. DETROIT - CHICAGO." On the left side of the receptacle "ALL STYLES AND SIZES FOR EVERY KIND OF FUEL." On the right "THE GENUINE ALL BEAR ABOVE TRADE MARK. BEWARE OF IMITATORS." The striker is underneath the match bin. Brightly colored in yellow, red, and black. Tin. 3 1/2" x 5". *From the Alperton collection*. $175-225.

A later version of the DeLaval Separator match holder. Note the number in use has risen to 1, 750,000. The message on the round piece at the top claims "Save $15 Per Cow Per Year." Painted black and blue with gold trim. American. Tin. 4" x 6 1/4". *From the Breon collection*. $200-350.

A cardboard match holder advertising "ACORN STOVES." On the ribbon is written "USED BY MILLIONS." The match container is made of tin and has a striking surface on the front. American. 4 5/8" x 6 1/2". *From the Breon collection*. $75-125.

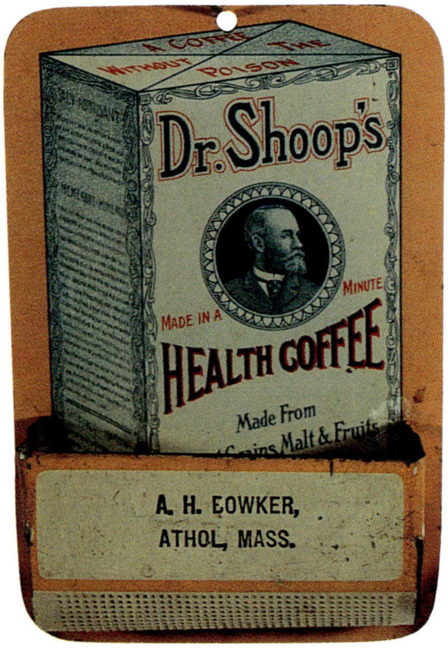

This advertising match holder reads "Dr. Shoop's Health Coffee, Made In A Minute, Made From Toasted Grains, Malts & Fruits, Price .25 cents. Made at the Dr. Shoop's Laboratories, Racine, Wis. U.S.A." On the front is the dealer's name and address, "A. H. Eowker, Athol, Mass." The top label reads "A Coffee Without the Poison." The striker is the pierced section below the match bin. Tin. American. 3 1/2" x 5". *From the Alperton collection*. $125-250.

This tin match holder advertises "RELIANCE BAKING POWDER Toronto, Can." The match container is a round miniature baking powder can. The background scene depicts a mother in the kitchen stirring up a batch of cookies while her daughter is standing by with a spoon ready to help. The advertisement at the top reads "Compliments of Ratcliff & Co. Stouffville, Ont." Canadian. 4 1/4" x 5 1/2". *From the Breon collection*. $300-500.

A tin match holder advertising "Her First BORN Steel Range." The scene depicts a smiling housewife removing a casserole from a Born oven. On the right side of the match bin is written "This reservoir is a luxury the year round." On the left side is "A little fuel cooks a large meal." Below is a ribbed area to be used as a striker. American. 3 3/8" x 4 7/8". *From the Breon collection*. $250-400.

This match holder contains an advertisement for "REX FLINTKOTE ROOFING For All Roofs Maxfield & Sanborn Hardware Pittsfield, N. H." The scene shows a barn with a horse in the stable. At the side of the barn is a grinder and an empty wagon. Tin. American. 3 1/2" x 5". *From the Alperton collection*. $150-250.

Match holder advertising "WILLIAM TELL FLOUR." The message displayed in the center reads "ANSTED & BURK CO. Roller Patent WILLIAM TELL Springfield, Ohio." On the left side is written "Makes Best Bread," and on the right side is "Biscuit And Pastry." The panel on the match bin reads "With Compliments of the Dealer Selling WILLIAM TELL FLOUR and Other Good Things." The striker is under the match bin. Tin. American. 3 3/8" x 4 7/8". *From the Breon collection*. $200-400.

Match holder designed for the Solarine Metal Polishing Company. The back panel features the product as well as the advertising slogan "WISE WIVES WORK WONDERS WITH." The message on the receptacle reads "POLISHES ALL METALS QUICKEST AND BEST." On both sides of the match bin "TESTED PROVEN BEST." The striker is on the lower front. Tin. American. 3 1/2" x 5". *From the Alperton collection*. $200-300.

Tin match holder for the wall showing a picture of the "DOCKASH STOVE FACTORY" in Scranton, Pennsylvania. The single container for the matches is on the lower front with a striker underneath. American. 3 3/8" x 4 7/8". *From the Breon collection*. $100-200.

A beautifully lithographed match holder to hang on the wall. The advertisement on the match bin reads "The Brown Shoe Co's STAR - FIVE - STAR $2.50 SHOES. For Men and Women." In the shield at the top is "The Brown Shoe Co. St. Louis U.S.A." Tin. American. 4 1/2" x 9". Rare, oversized. Price unavailable.

A match holder advertising "PHILIP MORRIS" cigarettes. Single compartment with a striker. Tin. American. 3 1/4" x 5 1/2". *Courtesy of Ken and Barb's Antiques*. $50-75.

Match Safes For The Pocket

Match safes for the pocket had the distinct advantage of being portable. They were manufactured in great quantities in every material available. Whether in plain tin or beautiful Art Nouveau sterling silver, these match safes are works of art and very collectible.

Although the advertisement on this pocket match safe is for mens' shoes, the front displays a beautiful woman admiring a rose. The body of the case is metal and the face is wrapped with celluloid. The top snaps open to expose the matches. The striking surface is on the bottom edge. American (c.1905). *From the Farriester collection*. $60-100.

View of above match safe showing the message on the back which reads "MADE SHOES FOR MEN FIT, WEAR AND SATISFY Nickerson - Bassett Shoe Co. 36 E. Colorado St. PASADENA, CALIF."

Match safe for the pocket with a steel body and a celluloid wrap. The front features a holly design with the message "Merry Christmas and Happy New Year." American. *From the Breon collection*. $75-125.

View of the reverse side of the pocket match safe to the left. This side reads "Compliments of HARRY AND BILL. H. W. Bean W. R. Stoner, Pool And Billiards, Cigars And Tobacco, Headquarters for all Sporting News. Phones Berkeley 1145 - 4927. 2130 Center Street, Berkeley, Cal."

A match safe for the pocket with a removable lid. Advertisement reads "FERDINAND WESTHEIMER & SONS DISTILLERS. Cincinnati, O. Louisville, Ky. St. Joseph, Mo. RED TOP RYE." Vulcanite. American. 1 1/8" x 2 3/4". *From the Breon collection*. $45-75.

A tin match safe for the pocket with a hinged lid. The advertisement reads "RED TOP RYE, America's Finest Whiskey." American. 1 5/8" x 2 5/8". *From the Breon collection*. $35-75.

Chapter Five

Silver Plated Match Holders

This special group of silver plated holders are shown in old catalogues as having a dual purpose, to hold either matches or toothpicks. For their design, whimsical subjects featuring children and animals were most favored. The quaint illustrations of children dressed in Victorian costumes designed by Kate Greenaway, an English illustrator of children's books , captured the hearts of Americans and were often depicted on the match holders. Fortunately, most of these containers have a company hallmark so they can be easily identified. A majority were made in the last quarter of the nineteenth century. They seldom had strikers.

A Kate Greenaway girl with outstretched arms is holding a hoop, which encircles a Venetian diamond amberina container. A dog is peeking around the right side. James W. Tufts #1146. Boston, Massachusetts. American. Base 2 3/8" x 3 1/2", 2 3/4" tall. *From the Aaronson collection*. $400-600.

A Kate Greenaway girl with outstretched arms stands beside a woven basket, which is the match container. The oddly shaped base stands on four ornate feet. Silver plate by James W. Tufts #2599. American. Base 2 1/2" x 3 1/2", 2 3/4" tall. *From the Aaronson collection*. $200-300.

A Kate Greenaway boy standing in front of a Venetian diamond amberina vase. Three ornate feet support the triangular base. Silver plate by James W. Tufts #2590. American. Base 3 1/4" x 3 1/4", 3 1/2" tall. *From the Aaronson collection*. $300-400.

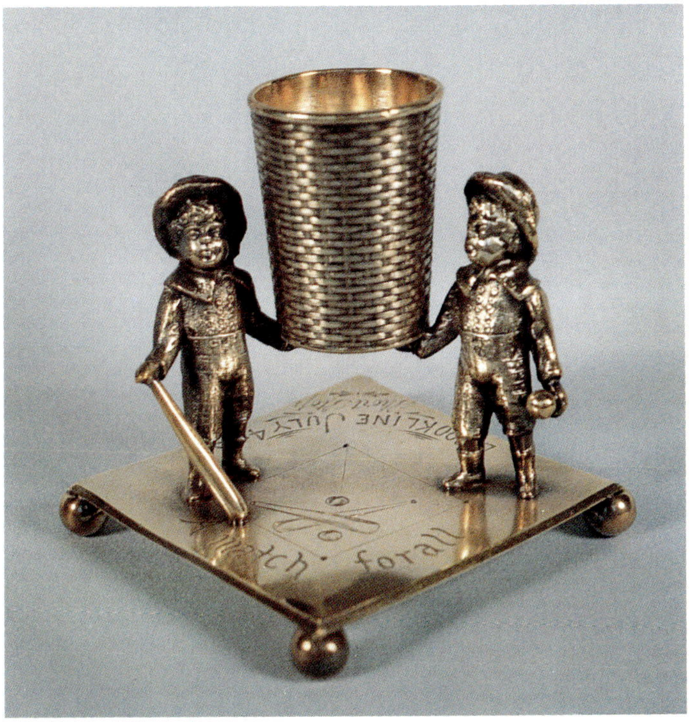

A charming match container with a Kate Greenaway design of two boys with a baseball and bat. The basket they are holding between them is for the matches. The base, on four ball feet, is a baseball diamond. The inscription on the base reads, "A Match For All. Brookline July 4' 93. Short Stop." Silver plate by James W. Tufts #2590. American. Base 3" x 3", 2 3/4" tall. *From the Aaronson collection*. $400-500.

A Kate Greenaway boy holding a cookie in one hand and a bucket in the other. The bucket holds the matches. Aurora Silver Plate Co. #355. Aurora, Illinois. American. Base 1 1/4" x 2 1/8". 3" tall. *From the Aaronson collection*. $150-200.

A Kate Greenaway type boy wearing a large hat is stretched out beside the match container. The receptacle is covered with a Repoussé floral design. Silver plate by James W. Tufts #3414. American. Base 2 1/2" x 2 3/4", 2 3/8" tall. *From the Aaronson collection*. $150-200.

Side view of match container on the left.

A small boy is removing his shoes and socks while leaning against a scalloped top match container. Derby Silver Co. #2307. Connecticut. American. 2" tall. *From the Aaronson collection*. $150-200.

A Kate Greenaway girl wearing a large bonnet sits on the left while a lamb stands to the right of this match urn. The front of the urn is decorated with a wreath. The oblong base rests on four plain feet. American. Base 3 1/4" x 4 7/8", 2 1/2" tall. *From the Aaronson collection*. $150-250.

A rabbit is crouching under a sprig of leaves and berries. The rough hollow log holds the matches. Silver plate by Simpson, Hall, Miller & Co. #33. Wallingford, Connecticut. American. The hollow log is 2 1/2" in diameter. 2" tall. *From the Aaronson collection*. $125-175.

A young girl is holding a beaded rope that encircles the match container. The container resembles a sack with a ruffled top. Silver plate by Rogers Smith and Company #44. American. 3" tall. *From the Aaronson collection*. $150-200.

A Kate Greenaway type girl is sitting beside a ruffled top match vase. American. Silver plate by Meriden Brittannia #47. *From the Aaronson collection*. $150-200.

A whimsical match holder featuring a squirrel with glass eyes blowing a horn. The ruffled topped container with a dimpled design holds the matches. Silver plate by Meriden Britannia Co., Meriden, Connecticut. #54. American. 2 1/4" tall. *From the Aaronson collection*. $150-200.

The figure of a sitting rabbit is the subject of this match container. A cup is encased in his back to hold the matches. Silver plate. American. 3 3/8" tall. *From the Aaronson collection*. $125-175.

A cat is standing on his hind legs with one paw on the goblet shaped match container. A pierced design encircles the top of the goblet. The rectangular base has a decorative border and is supported on foliated feet. Silver plate by James W. Tufts #3403. American. Base 2 3/4" X 3 3/4", 3 3/4" tall. *From the Aaronson collection*. $150-200.

A rabbit is sitting erect next to this match vase. An engraved design decorates the front of the ruffled top container. Marked "Meriden B #56." American. 3 1/2" tall. *From the Aaronson collection*. $125-175.

A small cat with glass eyes sits next to a barrel shaped match container. Silver plate by Wilcox #1909. American. 2" tall. *From the Aaronson collection*. $125-150.

A cat is standing on his hind legs peering into an egg shaped match cup. The egg sits on a wishbone and there are two small egg shaped feet in the front. Marked #39. Silver plate. American. 2 5/8" tall. *From the Aaronson collection*. $125-175.

A bear is scratching his back on this barrel shaped match holder. A decorative band encircles the middle section of the container. The round base has a rough surface on the top. Silver plate by Pairpoint Mfg. Co., New Bedford, Massachusetts. #3705. American. Base 3 1/4" in diameter, 2 5/8" tall. *From the Aaronson collection*. $125-150.

A small rat sits beside a sack of malt with the inscription "This Is The Rat That Ate The Malt." The round match container is ruffled and has a cord around the top. The circular base is scored and the edge is crimped. Silver plate. Pairpoint Mfg. Co. #3708. American (c.1885). Base 3 1/8" in diameter, 2" tall. *From the Aaronson collection*. $100-125.

A long tailed rat is crouching beside a match container with a ruffled top. The round match holder has crimped sides and sits on a flat bottom. Silver plate. Rogers Bros. #2303. American (c.1870). 2 1/4" tall. *From the Aaronson collection*. $100-125.

A violin is leaning on one side of this match vase and a top hat is on the other. The round ruffled topped container is covered with an ornate floral design. Meriden Silver Plate Co. #2514. American (c.1896). 2 1/2" tall. *From the Aaronson collection*. $75-100.

A tilted pitcher straddled by a wishbone sits on a triangular base. The message on the side of the pitcher reads "KIND WISHES." Silver plate by Simpson, Hall, Miller & Co. #0807. Wallingford, Connecticut. American.
2 1/2" tall. *From the Aaronson collection*. $50-75.

A square container resembling a sugar bowl holds the matches. Decorated with a floral and scroll design around the sides. Silver plate by Webster Co., North Attleboro, Mass. #330. American. 2 1/4" tall. *From the Aaronson collection*. $25-$45.

A beautifully detailed container. An elf is holding a large beer stein which serves as the match holder. The whimsical base has four feet in the shape of rutabaga with leaves. Silver plate marked "W.M.F." (Wurttembergische Metallwarenfabrik Geislingen, Germany). Base 4 1/8", 4 1/2" tall. *From the Aaronson collection*. $250-300.

A beautiful container with a floral Repoussé design and a scalloped top edge. Wilcox Silver Plate Co., #6103. American. 2 5/8" tall. *From the Aaronson collection*. $25-45.

A silver plated match holder with an open scalloped top. On one side is a lizard and on the other side a frog. Engraved around the sides is the message "I AM YOUR MATCH." Hallmark "E. G. Webster & Son N. Y." (c.1871). 2" tall. *From the Aaronson collection*. $90-110.

A ruffled top match container with a band of flowers around the sides. Wilcox Silver Plate Co. #2514. American. 2 1/2" tall. *From the Aaronson collection*. $25-45.

A small ruffled top match vase with a floral design and a plain oval center for an initial. Silver plate. Simpson, Hall, Miller & Co. #50. American. 2" tall. *From the Aaronson collection*. $25-45.

View of above match holder showing more of the engraved message.

A whimsical match holder featuring an ape dressed like a hunter with a dog by his side. The hollow log holds the matches and provides a hiding place for the rabbit in the hole at the bottom. Silver plate with the hallmark "James W. Tufts, Boston, Mass. 8400." Base 3 3/4" x 4", 3 3/4" tall. *From the Aaronson collection*. $500-600.

Kate Greenaway girl wearing a large hat is standing beside a Venetian diamond amberina match vase. James W. Tufts #3405. Ornate rectangular footed base 2 3/4" x 3 3/4", 5" tall. *From the Aaronson collection*. $225-275.

Left:
An oriental scene is the topic of this match holder for the table. A man is leaning against a barrel that holds the matches. A fan is displayed on the lower front. The square on the front of the barrel is for a monogram. Hallmark, "Simpson, Hall, Miller & Co." The round rimmed base is 2 3/4" in diameter. 4 1/2" tall. *From the Aaronson collection*. $375-425.

A Kate Greenaway girl stands on an ornate footed base next to an urn shaped match container. James W. Tufts #3404. Silver plate. Base 2 3/4" x 3 3/4", 4 3/4 tall. *From the Aaronson collection*. $175-225.

A silver plated match vase with a bear attached to the front. Marked "Queen City Silver Co. Cint'ti. O. " (c.1888). 1 3/4" tall. *From the Aaronson collection*. $100-150.

In the Victorian Era, after a dinner party, it was customary for the ladies to discretely retire to the parlor to discuss child rearing and household affairs. The gentlemen would sit around the table with an after dinner drink and a cigar discussing business and politics. The cigar stand was an elegant addition to the elaborate table settings that were customary in those days.

A baby chick with just its head exposed is breaking out of its shell. The broken egg cup holds the matches. On the top of the jagged shell is a corrugated plate for striking matches. Quadruple silver plate. Meriden B. Company #29. American (c.1879). 1 3/4" tall. $75-125.

Silver plated smoking stand with ten rings around the back to hold cigars. The tobacco box in the center has a removable lid and is flanked by two barrels holding matches. In front is an ash receiver. The stand rests on four bun feet and has a beaded decoration around the edge. Hallmark on the bottom reads "J G S" in a shield, and "E.P.N.S. 02637" which stands for electric plated nickel silver. Made in England (c.1890). 7 3/4" x 9 1/2", 5" tall. *Courtesy of Bondé Antiques*. $395-450.

Highly decorative cigar stand with a match holder. A deer is pulling a wheeled cart with a cupid riding on the top. The large bowl with a lovely bright cut design holds the cigars while the small egg shaped container at the top holds the matches. The stand rests on a rectangular base with four small ball feet. Marked "Wilcox S. Plate Co. 1784 Quadruple Plate, Meriden, Conn." (c.1885). Base 5" x 6 3/8", 7 3/4" tall. *From the Aaronson collection*. $1,400-1,600.

Chapter Six

PATENT ILLUSTRATIONS

"Pat'd.," "patd apld. for,"or other abbreviations for "Patent Applied For," can sometimes be found on match holders. This indicates the item was patented in the United States. Sometimes the date of patent and name of the manufacturer is inscribed on the article, but not often. When a patent number is found on a match holder the date of patent can be determined by looking at the patent table which appears near the index of this book. Unfortunately, most match holders are not marked with the maker or with the date of manufacture. Often, it is difficult, if not impossible, to ascertain where or when a match holder was made.

We have researched United States Patent Libraries and found some interesting examples which are well worth studying. The creative manner in which the inventor tackled the problem of creating a safe place to store volatile matches is instructive. The inventor would apply for a patent with a very detailed description of his match container. We have condensed this information, but kept the inventor's description. The patent numbers and dates of application are also included for those interested in further research. The United States Patent Office issued many patents for the "Match-box," the "Match-holder," and the "Match-safe," all terms used to describe items that have the same purpose—a safe and convenient way to store matches.

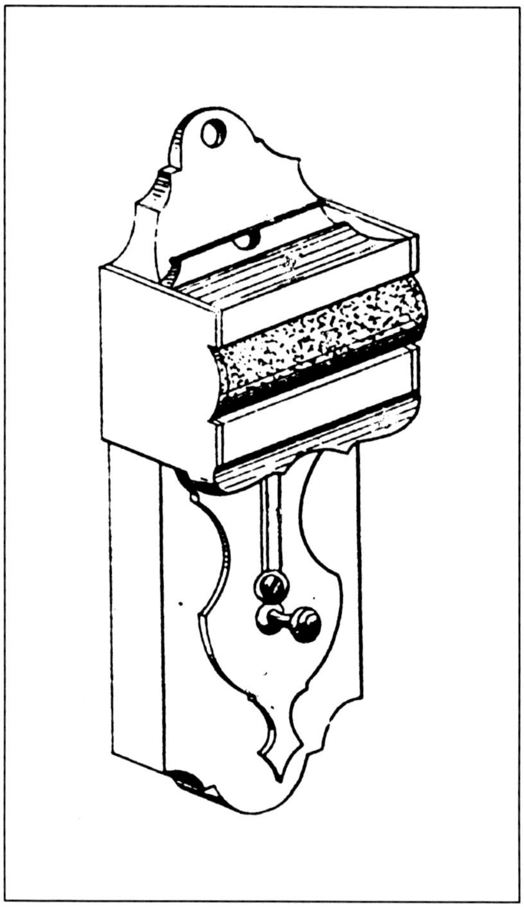

This invention relates to a class of match safes designed to deliver a single match at a time. The novelty exists in the simple slide and hopper arrangement. A single match will lodge in the groove on the slide, filling it so as to prevent others from lodging when pushed up and out through the upper open space. Thus, upon repeating the motion of the slide to its full extent, match after match will be delivered, until the hopper is empty. The outer space of the head portion is made ornamental with an intermediate space for a strip of sandpaper on which to ignite the match.
Pat. No. 149754. Patented by J. S. Henry in 1874.

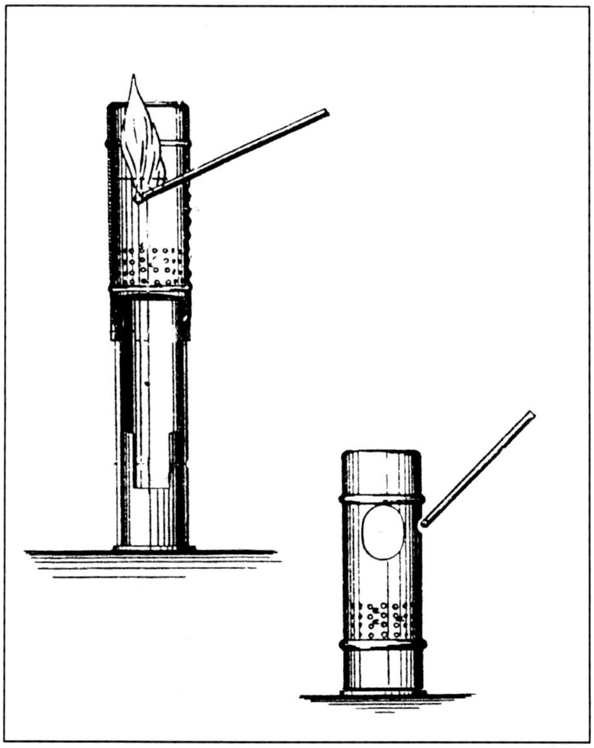

Combination matchbox and cigar-lighter. Matches are introduced into the box through the slot, and in this way the box is filled. When it is desired to light a match under circumstances where it would probably be extinguished by wind or a draft of air, the guard is slid upward, and a match is taken from the box through the slot. The head of the match is rubbed on the corrugations, which ignites it. The ignited end is then slipped into the opening within the guard. The perforations carry the flame of the burning match above and clear from the top of the guard, where it may be used for lighting purposes. When the matchbox and lighter are not in use, the guard is slipped over the box. This effectually closes the slot, keeping the matches safely within the box.
Pat. No. 148411. Patented by G .J. Capewell in 1874.

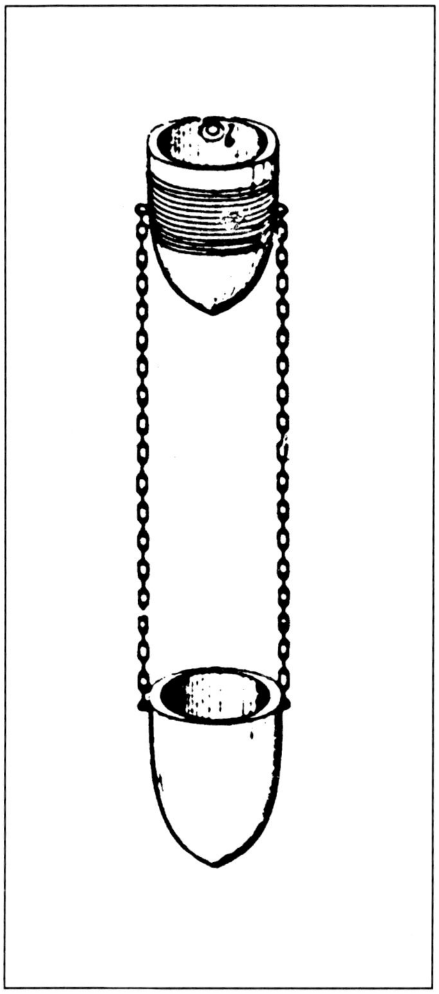

Two cup match holder connected by chains. The upper cup is grooved or toothed on its outer surface to enable a friction-match to be inflamed by being rubbed smartly against it. The top cup holds friction matches and the lower one holds burned matches. By being suspended by chains, one of the cups can easily be turned down and emptied at any time independently of the other.
Pat. No. 157531. Patented by H. H. McBurney in 1874.

This receptacle for matches, or other articles, in which two receptacles are so connected that by opening one the other is pushed out, and thus two are opened simultaneously, consists in connecting a segmental cover in such a manner with the lower drawer and receptacle that, in pushing back the cover, the lower receptacle or drawer is pushed out. The cover has a roughened surface on top, which is presented when closed, and provides a place to rub the match. One side is for matches and the other for burned matches.
Pat. No. 173195. Patented by J. G. Warren in 1876.

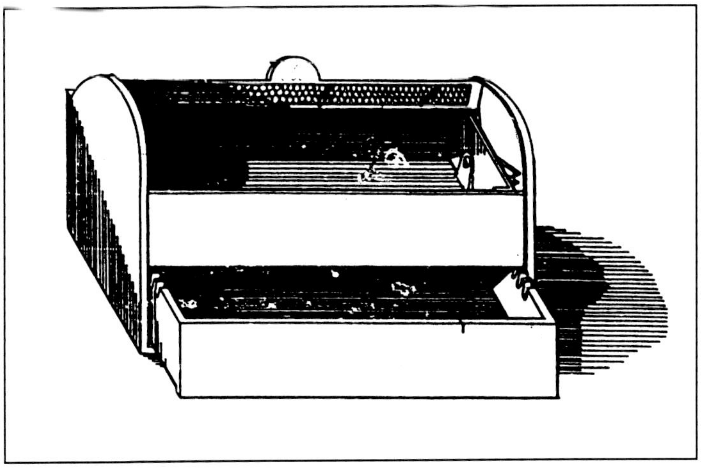

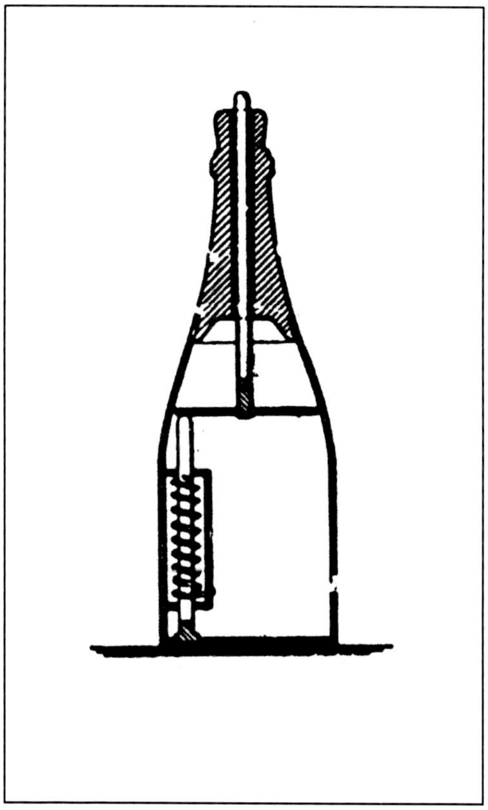

This match safe is made in the general form of a bottle. The neck and stopper are of cast metal and the body of sheet metal. There is a push pin running through the neck of the bottle, connected to a disk within the hollow sheet metal portion. There is a movable cover with a spring to keep it shut. When the push pin is pressed upon, its disk acts against the fulcrum pin, pushing the cover out of the match safe; its spring turns it aside. The matches will be placed in the case and the cover closed. The safe will usually stand upon the bottom of the bottle. Hence, in opening the safe, it should be held bottom up. By pushing on the pin, the disk and pivot wire will be moved and the cover raised .
Pat. No. 195928. Patented by W. L. Hubbell & F. Ochs in 1877.

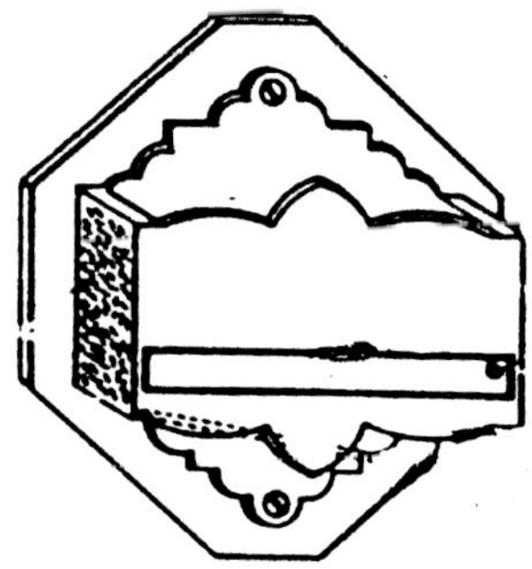

Fig:1.

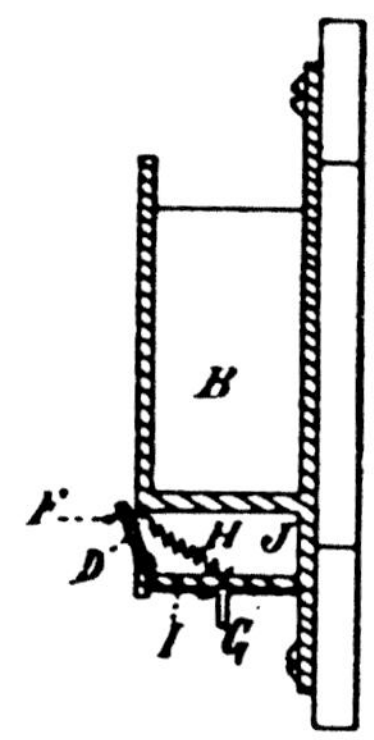

Fig:2.

This receptacle is provided with a door, having a thumb piece or pull, which is hinged at its lower edge to open outwardly, and a spring, operating contractively, to keep the door closed when not in use. The bottom of the receptacle is also hinged at one end and provided with a button or clasp for retaining it in position when the receptacle is filled. The matches may be ignited by striking them on the roughened plate and the stubs deposited in the receptacle. The door can be opened by means of the thumb piece for that purpose.
Pat. No. 179202. Patented by H. Kinsley in 1876.

Combined match safe, pin cushion, and stamp box. This article for the pocket consists of a match safe on half of one of the sides and a pin cushion on the other half of the same side. A receptacle for postage stamps is on the other side.
Pat. No. 179897. Patented by I. C. Cowles in 1876.

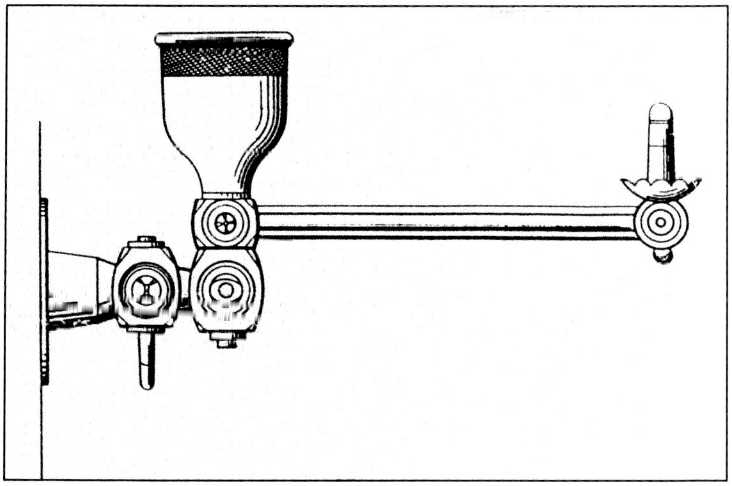

Gas bracket-match safe. The striker or roughened plate on which the matches are made to ignite by friction can be replaced. The match holder and striker are removable from the bracket for ease of cleaning.
Pat. No. 150892. Patented by G.W. Righter on May 12, 1874.

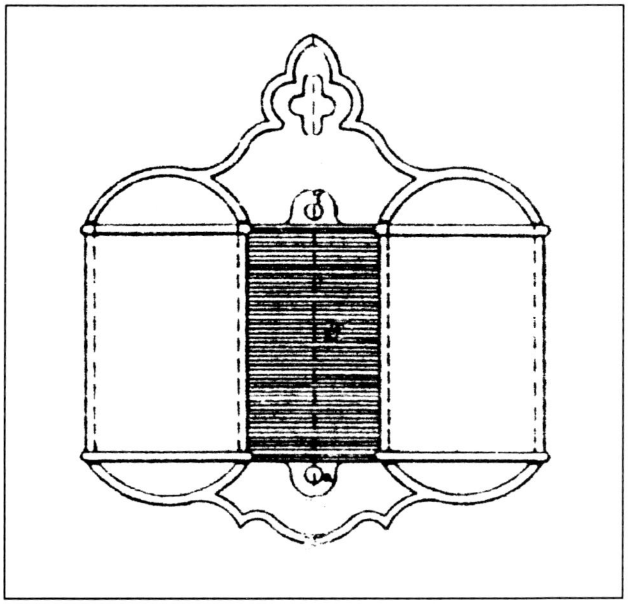

Match holder. The striker is the roughened plate attached to and between the pockets in an inclined position. By placing the striker in the inclined position, less force is required to ignite the match. The ignited match is also sufficiently far from the wall to prevent the wall from being discolored by smoke and flame. The pockets, located on either side of the striker, serve to guide the match to the proper place for striking.
Pat. No. 177837. Patented by Oliver B. Hicks on May 23, 1876.

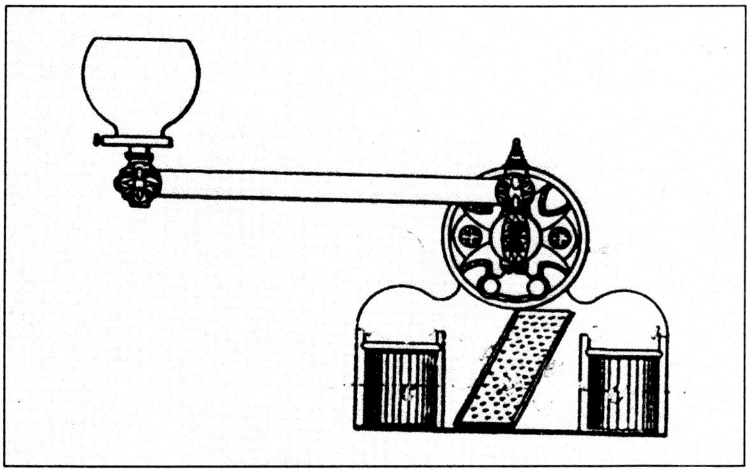

This invention consists of an attachment to a gas fixture, which constitutes the match safe or holder, wherein the matches may be readily held and are conveniently accessible.
Pat. No. 180995. Patented by W. Ferguson on August 15, 1876.

The object of the invention is in a ready, simple, feasible, and thoroughly practical manner. Solely by the movement of the holder, which effects the lifting of a match from the interior of the holder to the exterior thereof, the match is tipped and caused to occupy a vertical position. The match tipping device is secured to the under side of the cover and projects slightly beyond the terminal of the slot. By this arrangement, when the holder is down the match will be held in a vertical position by being clamped between the tipping device and the vertical wall of the recess. All danger of igniting the match when removing it from the holder is eliminated. Pat. No. 740710. Patented by J. J. Sullivan on October 6, 1903.

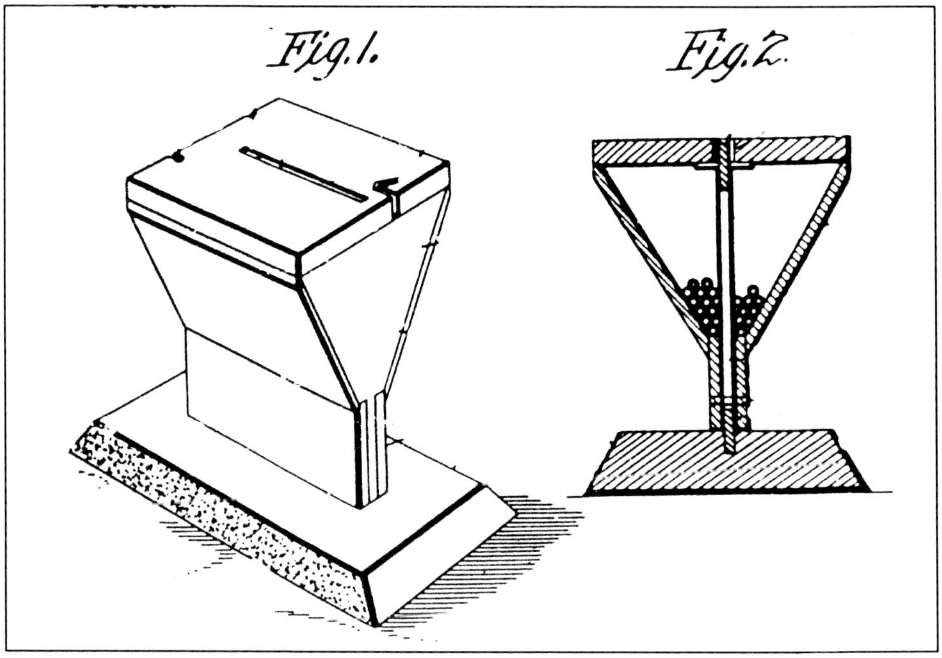

This invention belongs to the class of inventions designated as "match safes or receptacles." It is designed for holding matches in quantity and admitting one at a time. Thereby it places a check upon the practice of taking a number of matches at one time, as in receptacles in which the matches are readily accessible. The trough is closed at one end and is open at the opposite end for the ready discharge of the matches one at a time. In order to accommodate the safe to the varying sizes of the matches, a wall is provided at one side of the box and is adjustable toward and away from a side wall. It is held in an adjusted position by friction.
Pat. No. 728027. Patented by Jonas W. Selander on May 12, 1903.

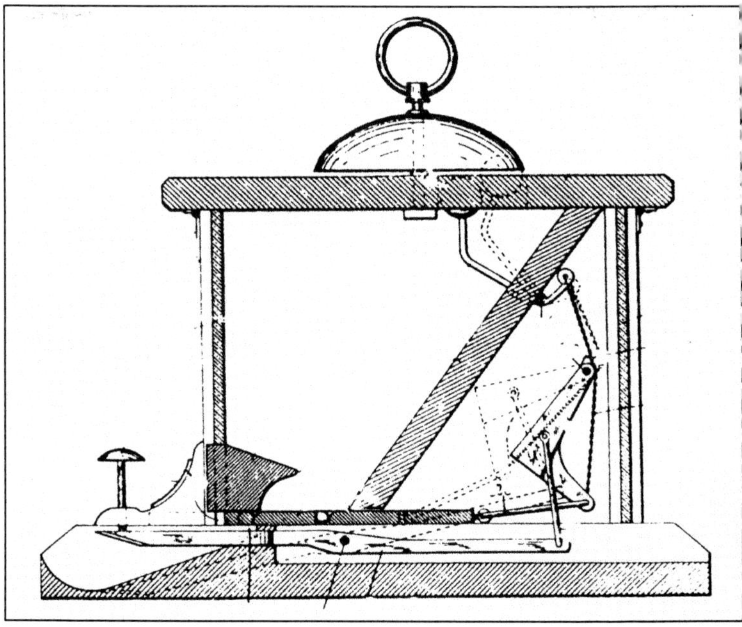

The object of this match box is to provide a strong, durable, and neat construction to deliver matches singly when the operative lever is pressed while at the same time ringing an alarm bell and thus saving waste of matches. In shops, hotels, and the like, it makes an amusing toy for the patrons.
Pat. No. 731785. Patented by F. K. Kennedy on June 23, 1903.

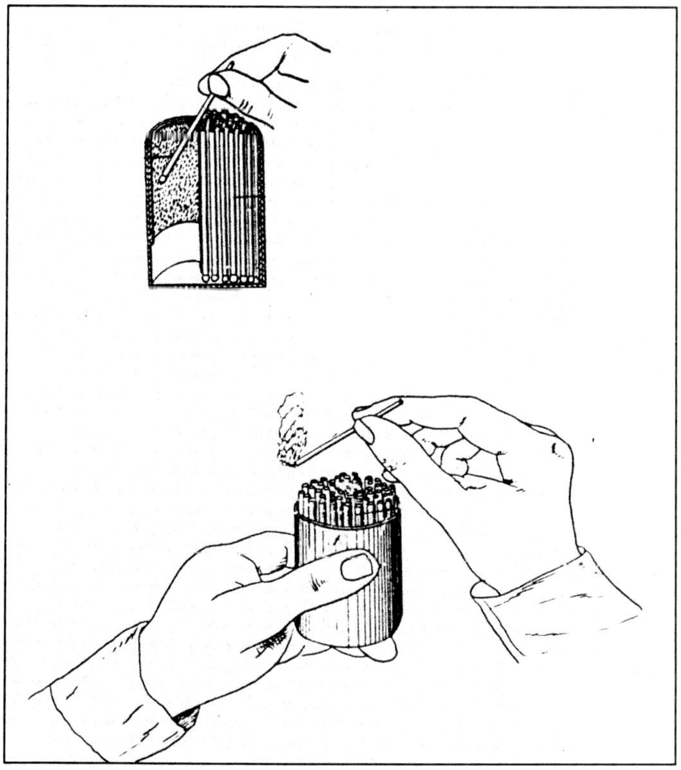

This invention lines the interior surface of a match box with sandpaper within about a half of an inch of its bottom. Then, when the matches are placed within the box with igniting ends downward, their composition ends will not be in contact with any frictional surface. When they are withdrawn, their igniting ends are brought in contact with the roughened surface on the interior of the box and thereby ignited.
Pat. No. 32187. Patented by S. W. Francis on April 30, 1861.

This invention consists of a combined lamp bracket and match safe. This is a novel arrangement on the same bracket back of sockets for lamp brackets and a match safe. Matches are always kept in proximity to the lamps, just where they are wanted, and the inconvenience of running about to obtain a match whenever one is needed is obviated.
Pat. No. 141581. Patented by A. Meyersberg on August 5, 1873.

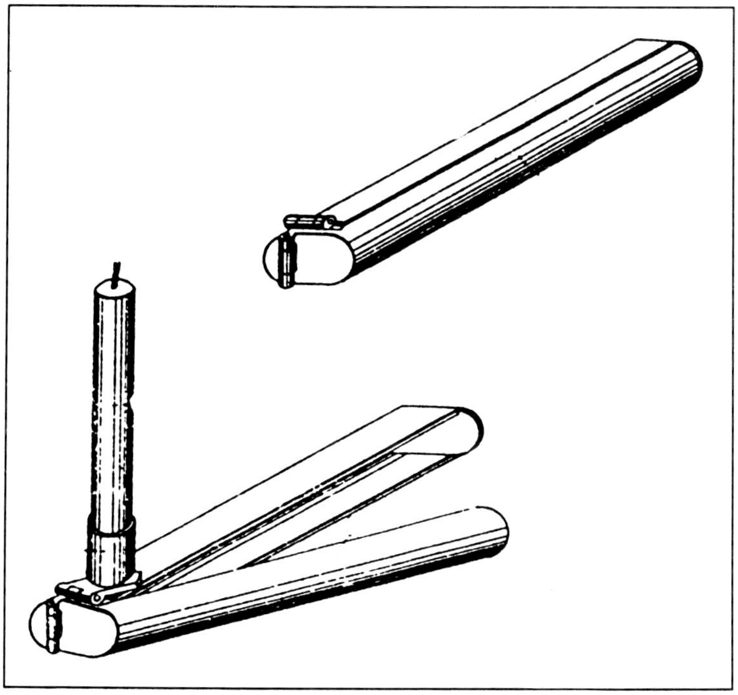

The object of this invention is to provide a convenient means for furnishing a light to be used on occasions when an ordinary lamp may not be easily accessible or convenient to carry about. This invention consists of a folding case of a size convenient to be carried in the pocket, and provides compartments for holding matches and a taper or candle. The latter may be readily lighted and carried about, or with its holder, it may be placed upon a support for the purpose of a lamp.
Pat. No. 65142. Patented by John A. Whipper on May 28, 1867.

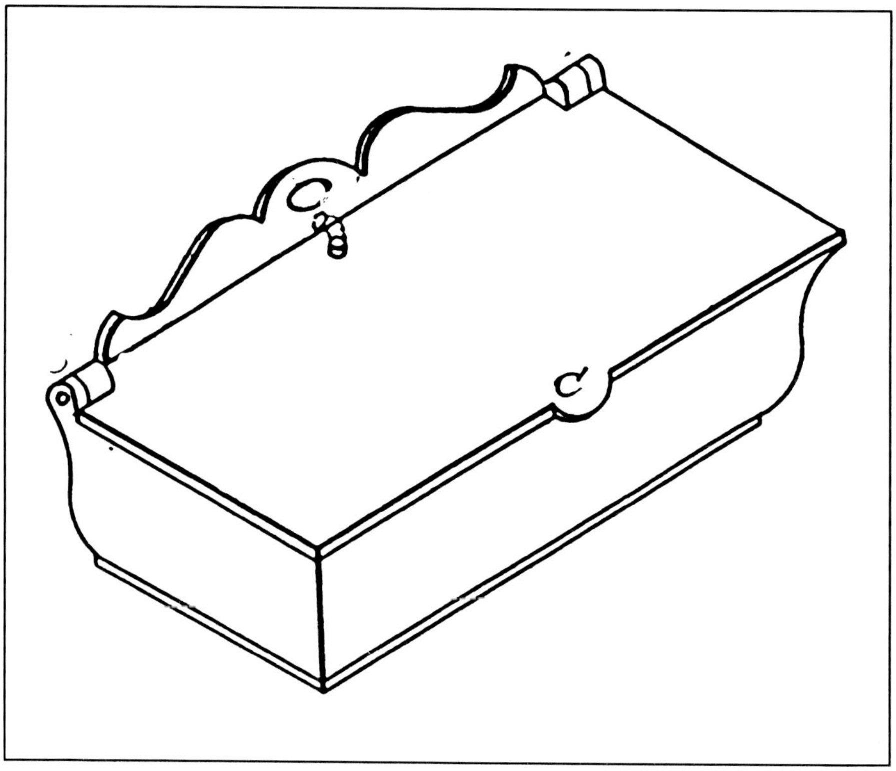

This invention relates to an improvement in boxes used for containing a supply of friction-matches, commonly called "match-safes." On the upper side of the lid, near the back edge, there is a projection of sufficient height to strike the back of the box before the lid is raised to the perpendicular. At that position the projection strikes the back, so that the lid cannot be raised higher. Consequently, when the hand that raised the lid is removed, the lid will fall as a result of its own weight, so that the box is not left open to expose the contents
Pat. No. 45554. Patented by Wm. H. Andrew on December 20, 1864.

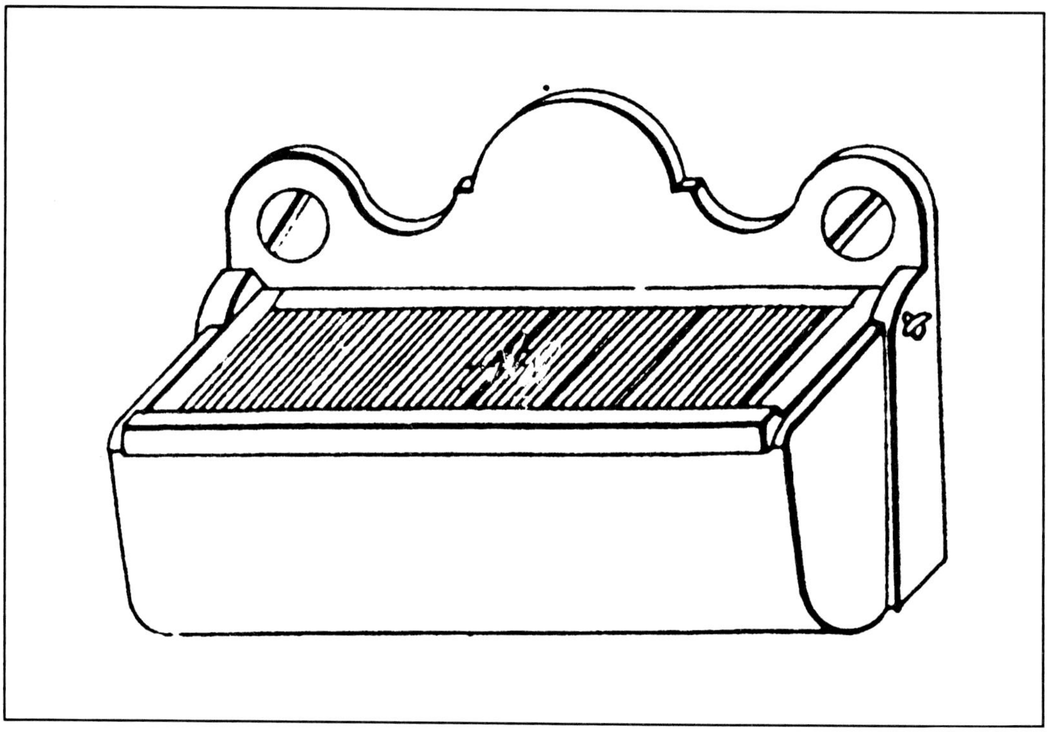

The nature if this invention is to provide a match-safe with a metal grate to enable one to kindle friction matches with facility. It is less liable to wear out like other match lighters.
Patent No. 76786. Patentee, Alexander Martin on April 14, 1868.
Note: Not all patentees were Americans. This patent is from Switzerland.

The object of this invention is to provide a safe and convenient receptacle for matches that can be connected to a gas burner bracket. The bracket pipe, which projects from the wall, carries the gas burner. The match safes are on each side of the burner. Match safes attached to the bracket plate cannot get lost, stolen, or mislaid. Instead, they will always be found just where they are most convenient.
Pat. No. 144283. Patented by M. L. Orum on June 21, 1873.

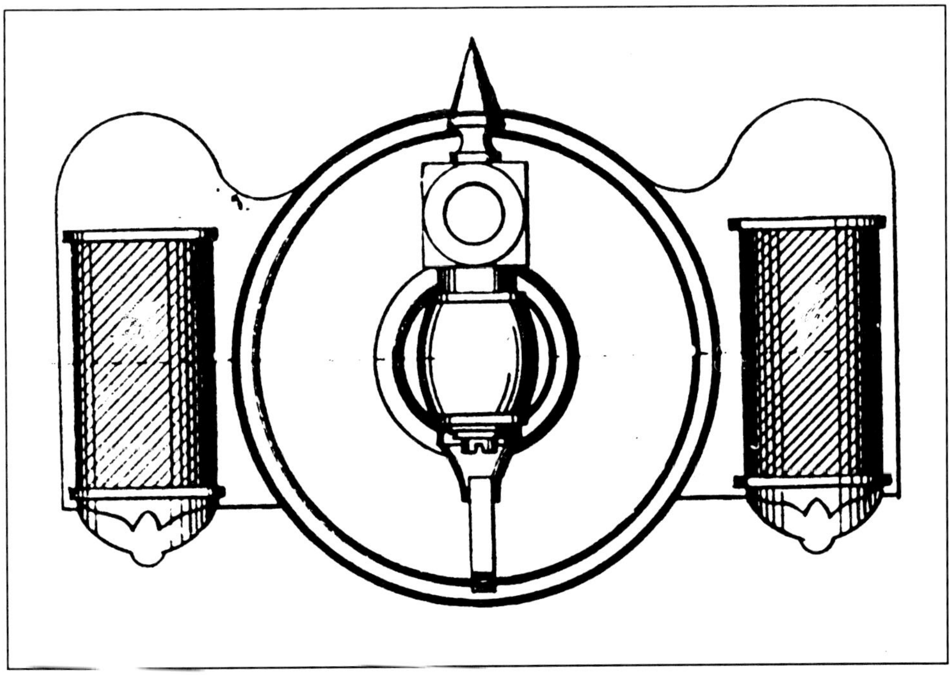

The object of this invention is to afford the means for easily attaching the friction paper and preventing it from being seen when not required for use. It is well known that the friction surface of paper becomes soiled by use. This materially injures the appearance of the safe. This invention is designed to obviate this objection. The match box is attached to an ornamental back plate connected thereto by a hinge and provided with a convenient means for holding the friction paper and retaining the match box plate in place when folded against it.
Pat. No. 188250. Patented by J. A. Kratt on March 13, 1877.

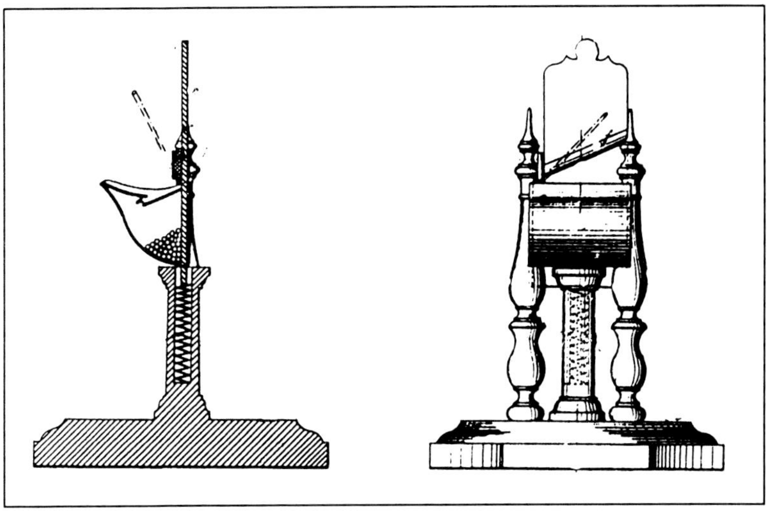

The nature of this invention consists in a match box that is reliable, safe, and self acting. The object is to furnish a cheap match box that may be placed in any convenient place, and one that places one match at a time in position for lighting without fail. It also will prevent children from being injured by matches. This case is provided at its upper side with a small opening through which the matches are introduced into the case. They need only be placed on the inclines; then they will roll in to the case. Above the spring and sliding in grooves in the columns is a slide, having in its front an inclined groove into which the matches fall. This brings the matches above the top of the case, stands them in an inclined position ready to be taken in the hand, and strikes them against the rough surface of the striking plate.
Pat. No. 153081. Patented by L. Jarchow on July 14, 1874.

Shown in the open position to permit the insertion of a match box from below, this invention relates to improvements in a combined match box holder and safe. To insert the match box, it is only necessary to swing the body of the device upwardly to a horizontal position. The box of matches, open at the top, may then be placed in position. After which the body of the device is lowered and some of the matches will pass into the auxiliary chamber, which will be kept replenished until all the matches in the box are used. The body of the device will then be raised, the old box removed, and a full box substituted.
Pat. No. 739273. Patented by C. F. Paige & S. H. Ullrich on September 15, 1903.

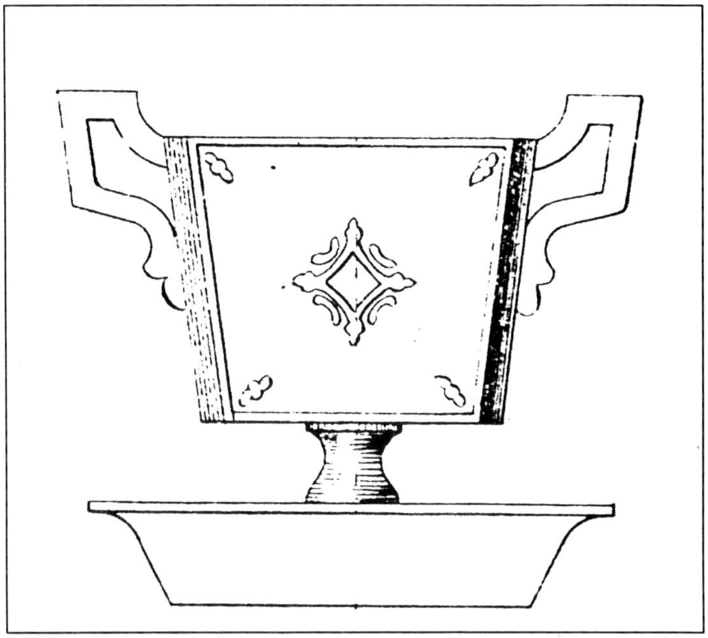

This invention relates to an improvement in the construction of receptacles for matches. The object was to cast a receptacle of ornamental shape. The receptacle represented here is in the form of a vase, in two parts, divided through the center; the division extends through the base. The herein described matchbox is secured to the base plate by a tenon of dovetail form.
Pat. No. 121788. Patented by Albert. D. Judd. on December 12, 1871.

This invention relates to an improvement in the class of matchboxes that are provided with a swinging or hinged cover. It exists in the peculiar manner of forming the hinged connection between the box and cover. As a convenience for setting the box upon a shelf or in any other position, the front legs correspond in height to the lugs in the rear.
Pat. No. 122836. Patented by Albert D. Judd on January 16, 1872.

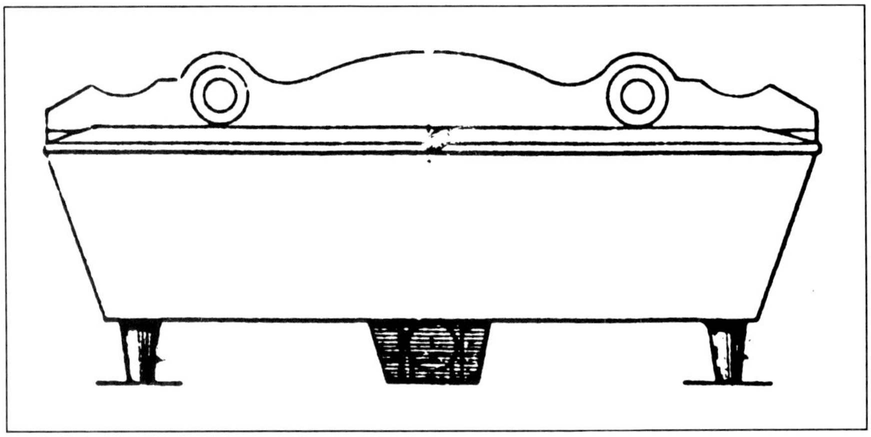

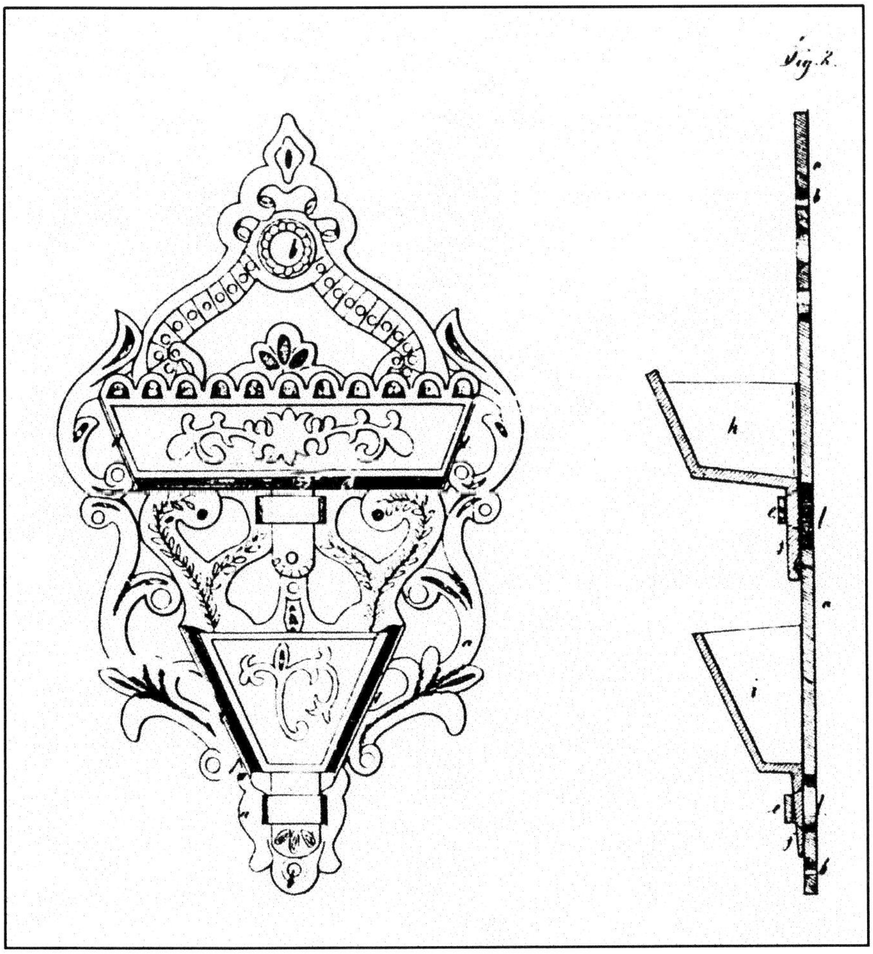

This invention relates to that class of match-safes in which the box or safe is attached to a plate that forms the back of the box. It also has a plate for setting it against the wall. The upper box is to be used for matches, and the lower box is for the burned ends. By placing one box below the other, the boxes are of sufficient length to allow the matches to lie horizontally; this is a great convenience. There is no risk of taking a burnt match instead of a fresh one, as is frequently the case where the matches are in boxes placed side by side.
Pat. No. 102828. Patented by Albert D. Judd on May 10, 1870.

This safe is designed to furnish a safe place for keeping packages of matches as well as those separate for daily use. The packages are placed in the large box or receptacle, while those separated for daily use are placed in the delivery box. Access can more easily and conveniently be had to the separated matches when held in their receptacle on the outside of the box, which would not be the case were they contained within the main box. It would then be necessary to raise the heavy cover of the box each time it was desired to take a match.
Pat. No. 134523. Patented by L. Darozir on January 7, 1873.

Left:
This invention consists of a match safe with a delivery aperture at the top, of peculiar shape, adapted for the introduction of the thumb and finger. This permits the withdrawal of the matches only one at a time. In the accompanying drawings is a perspective view of the improved match safe in a form adapted for setting upon a table, counter, or shelf. This view hereof is in a form adapted for hanging on a wall. If a number of matches are placed on top of the box, a slight shaking or jarring of the box will cause them to descend quickly through the aperture.
Pat. No. 144615. Patentee J. & A. Helm. Patented on November 18, 1873.

Below:
This invention differs from others by providing, in the central part of the box top, a longitudinal slot with recesses communicating with it on both sides. In this way, not a finger merely, but both the thumb and finger may be introduced, one on each side of the slot, and a single match picked up and withdrawn without pressing or rolling in upon the other matches or against the side of the box. The invention on this account possesses superior safety, especially when used with what are known as parlor matches or other matches tipped with an explosive composition.
Pat. No. 144615. Patented by Jules Helm & Alphonse Helm on November 18, 1873.

This invention makes certain improvements in Boxes for Containing Matches. Hitherto in affixing matchboxes in dwellings, it was almost , if not entirely, impossible to prevent the defacement of the wall or partition by the nail or other fastening device from which the box hung. This box can be easily secured at the side of any window-frame or doorjamb without being fastened to the wall. Pat. No. 124886. Patented by Theodore Crommelin on March 26, 1872.

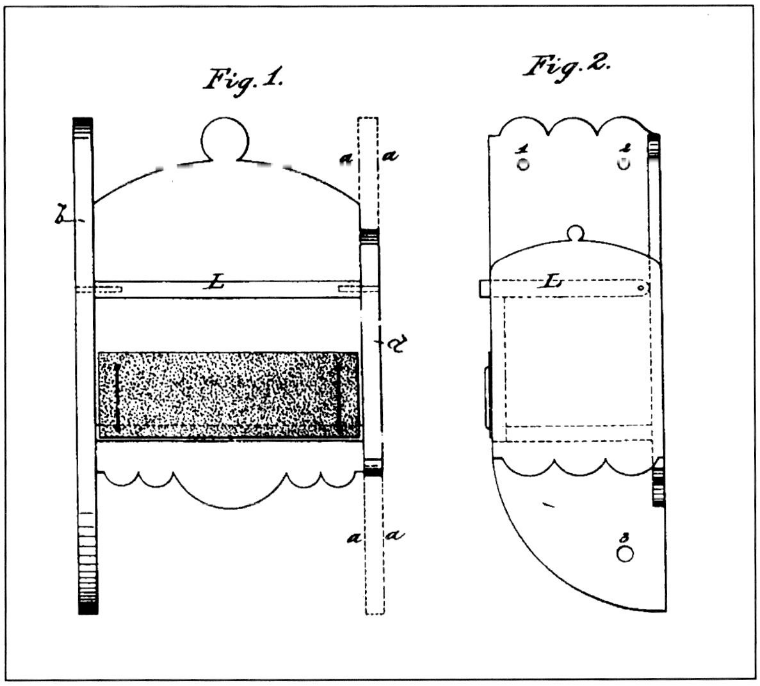

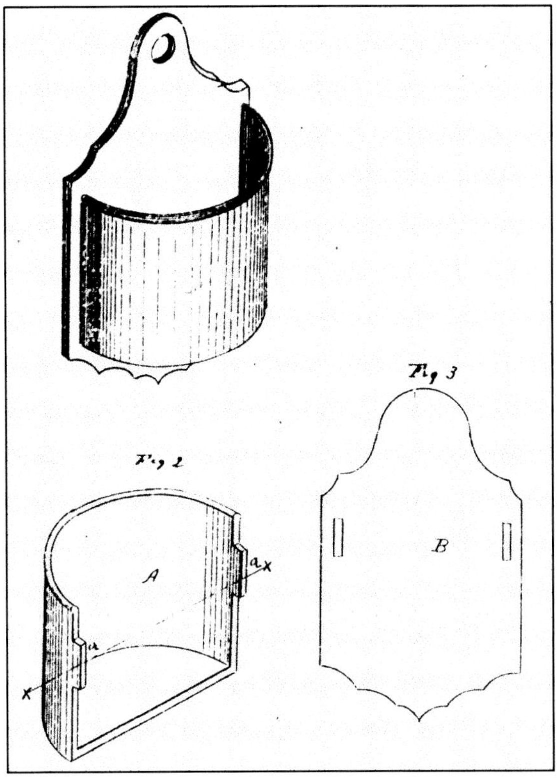

This invention relates to an improvement in the construction of a receptacle for matches. It is designed to be hung or secured to the wall and consists in a base, or back, and a front; both are formed from cast metal. The front was cast with lugs and placed in a mould, and then the back or base was cast thereon, attaching itself to the front during the process of casting and by means of the lugs on the front. Thus, a receptacle was formed in the simplest and cheapest possible manner, and a neat and durable match safe was created.
Pat. No. 73706. Patentee Bradley and Hubbard on January 28, 1868.

The upper compartment or reservoir of this safe forms a receptacle in which the matches are placed and prevented from falling into the lower compartment by the back of the lid or cover. The lid is held in its place by the trunnions to which it hinges with the body of the box. Longitudinally with the trunnions, the back part of the lid is formed in a semi-cylinder. Into this a small recess or groove of sufficient size to contain but one match at a time is made. When the lid is down, the recess is in such a position that one of the matches in the upper compartment will fall into it. By raising the lid, the match in the recess or groove falls backward and into the upper part of the compartment.
Pat. No. 102807. Patented by M. Russell on May 10, 1870.

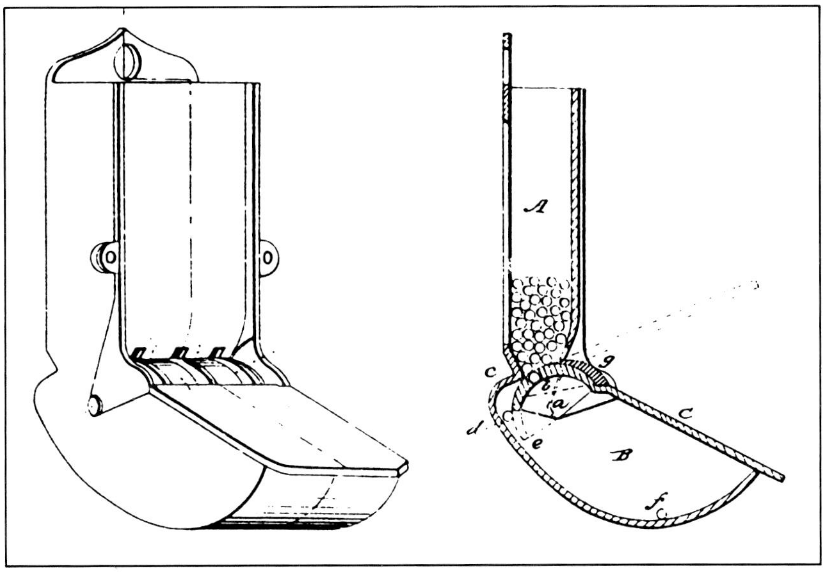

This invention relates to a match-safe combined with a drawer or receptacle in which the stubs or waste ends of the matches may be deposited after the matches have been ignited and used. The safe consists of two pockets resting upon a base. The plate forms the back of the pockets and extends upward as a means for suspending the device from a nail driven into the wall. The base is a box open at the ends with a drawer inside. The front of the base has a slot cut lengthwise thereof. When the drawer is closed, the stubs or waste ends of the matches can be passed therein.
Pat. No. 141793. Patentee, John Holtz on June 11, 1873.

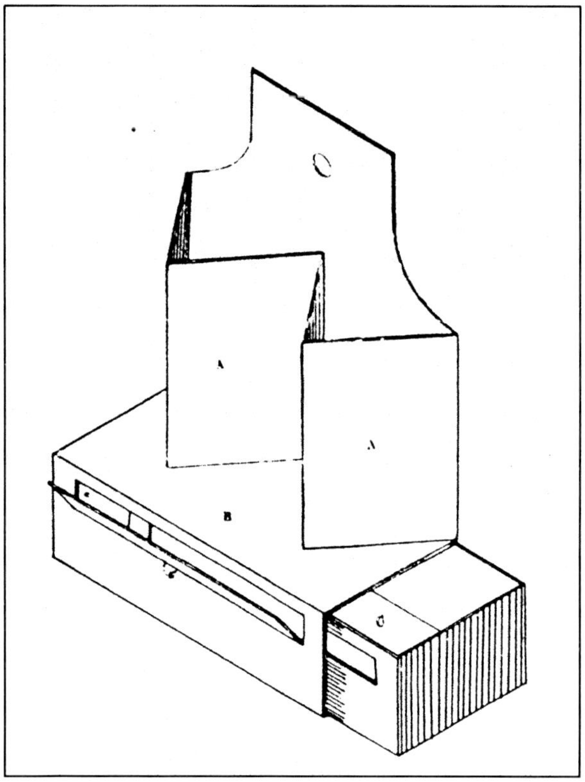

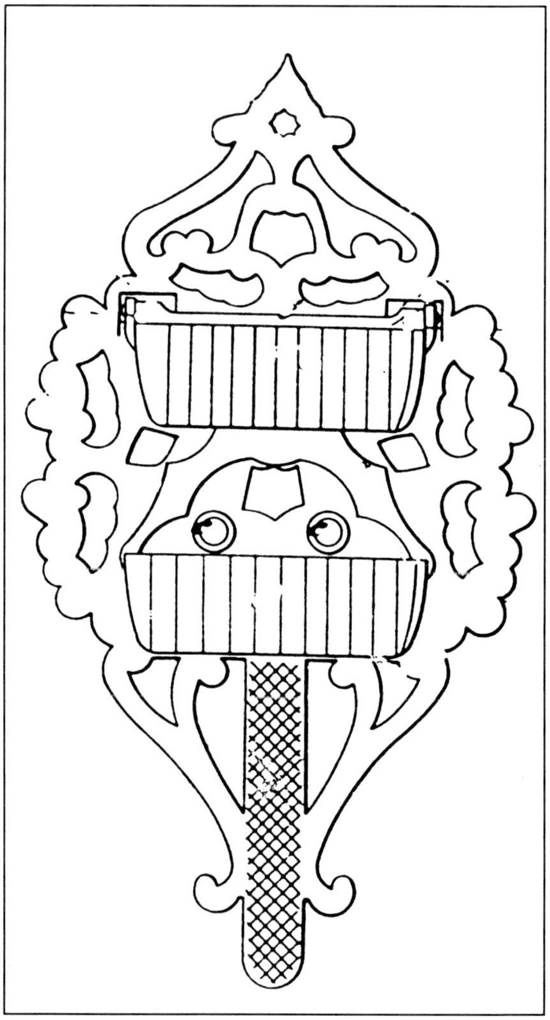

The nature of this invention consists in producing a whole device or safe complete in three pieces of metal, and further in securing the whole together by simply heading the rivets. There is the bracket or plate to which the boxes and lid are secured, with screw holes for securing the device on the wall and a rough friction-surface for igniting the match, the match receptacle with lid, and the waste box.
Pat. No. 79426. Patented by H. M. Woodford on June 30, 1868.

This invention relates to an improvement in the manner of securing together the parts of a cast metal match-safe. The box and back have been secured together by means of rivets, an expensive operation, and one liable to break the parts. To overcome these difficulties is the object of this invention. Ears were formed upon the front edge of the box part which correspond to recesses in the rear. By slightly compressing the two sides, the ears on the front spring into the corresponding recesses in the rear, thus securing the two parts together. By this construction all liability to break, as in the process of riveting, is avoided, and the expense of putting the parts together is comparatively nothing.
Pat. No. 102676. Patented by G. Geer, assignor to Charles Parker, on May 3, 1870.

The object of this invention is to create a match-safe or box that will admit of the matches being removed, or taken therefrom, with greater facility than usual and without the liability of spilling or scattering any out from the box while in the act of withdrawing one or more from the same. In withdrawing the matches from the box, all that is required is simply to pass the thumb and finger through the opening and raise them out of the box. The lid is raised by the upward movement of the thumb and finger, and is closed by its own gravity as the thumb and finger are drawn from the box. The lower part of the plate below the box may be corrugated, as shown, for the purpose of rubbing the matches upon it to ignite them. Pat. No. 82752. Patented by H. Richmond, assignor to Charles Parker, on October 6,1868.

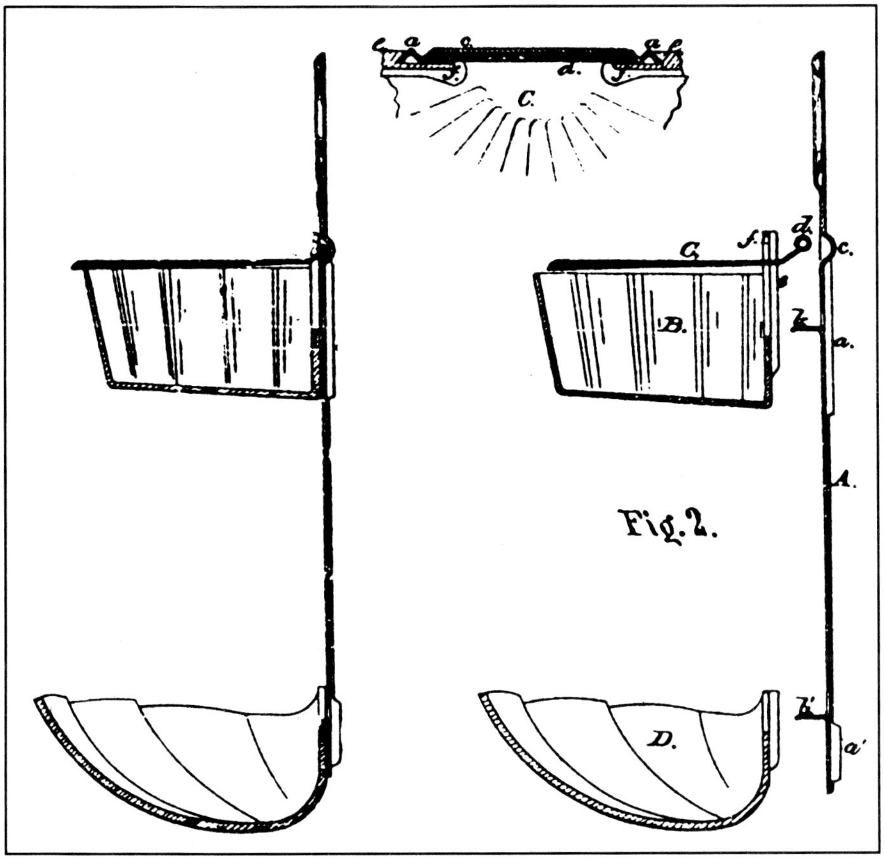

This invention relates to the safes which hold matches. The match-safe is composed of four pieces: namely, the back, the match-box, its cover, and the shell for receiving the burnt matches. The back plate is provided with a hole under the match-receptacle for the purpose of securing the safe against the wall. The sheet metal parts are corrugated providing a roughened surface on which to ignite the matches.
Pat. No. 138261. Patented by Clinton H. Geggett on April 29, 1873.

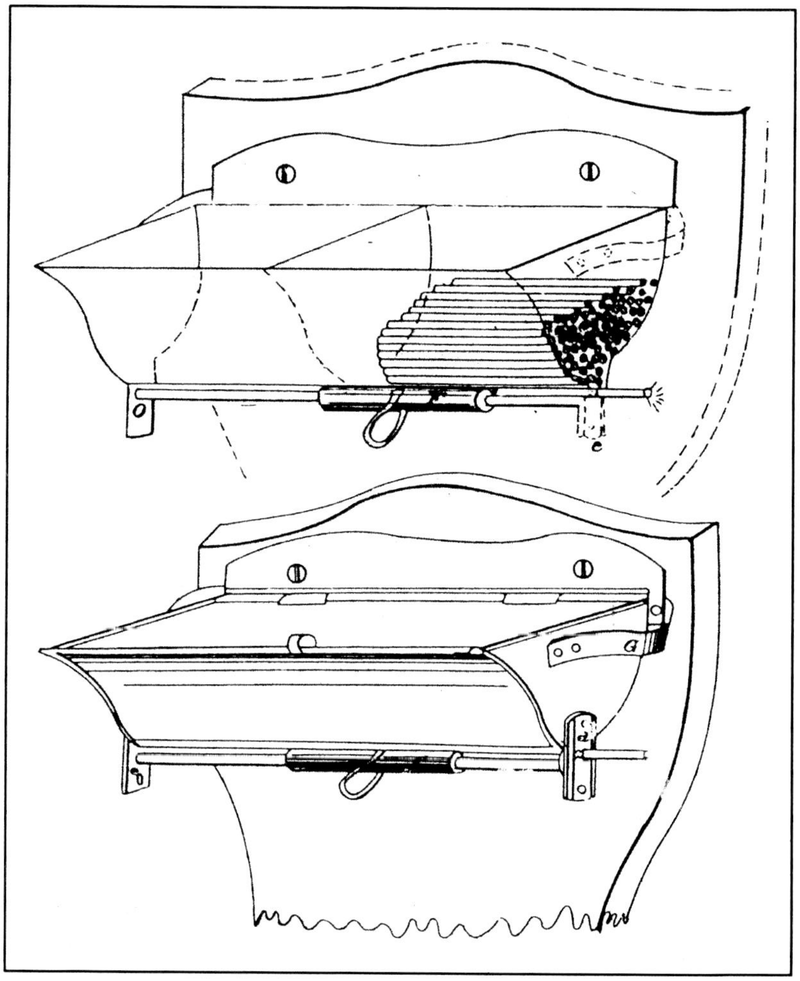

This invention makes a match-safe a little longer than twice the length of a common round match, has a partition dividing it in the middle, and narrows at the lower or bottom ends. The bottom is left open, and a rod and slide are attached below for the purpose of thrusting the matches from the safe through openings on the sides provided with sharp-toothed spring-plates. These scratch and ignite the match.
Pat. No. 70362. Patented by Adolph Roesler on October 29, 1867.

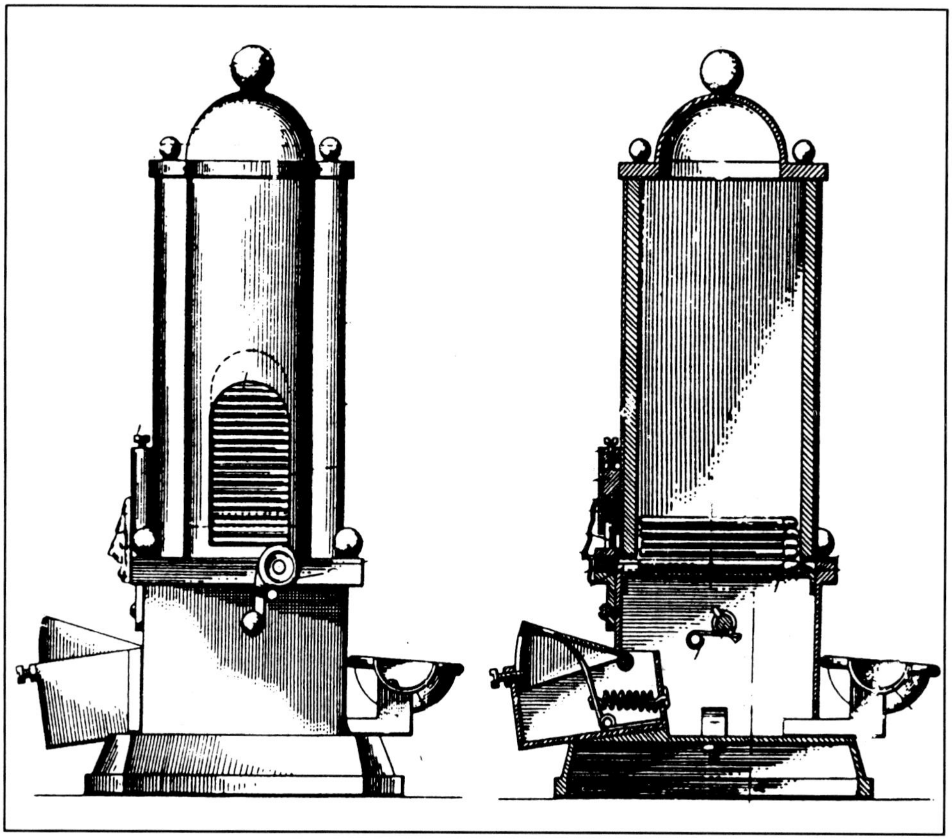

This invention relates especially to devices employed for cutting off the tips of cigars before lighting them and provides matches for igniting the same. It has for its object the provision of a simple, cheap, effective, and ornamental device. There is an opening for the egress of the matches, and at the mouth of this opening is a scratching device in which there is a fixed semicircular serrated block. The magazine is filled with matches. The operation of the device is as follows. By turning the crank to the left, the arm comes in contact with the head of the lower most match; this forces the match partially out of the opening. The match is then drawn out between the finger of the user, the head passed between the serrated blocks, and ignited.
Pat. No. 730841. Patented by G. Prioletta on June 9, 1903.

Table 1
United States Invention Patents And Their Time Frames

Year	Number	Year	Number	Year	Number	Year	Number
1836	1-109	1877	185,813-198,732	1918	1,251,458-1,290,026	1959	2,866,973-2,919,442
1837	110-545	1878	198,733-211,077	1919	1,290,027-1,326,989	1960	2,919,443-2,966,680
1838	546-1,060	1879	211,078-233,210	1920	1,326,890-1,364,062	1961	2,966,681-3,013,102
1839	1,061-1,464	1880	233,211-236,136	1921	1,364,063-1,401,947	1962	3,013,103-3,070,800
1840	1,465-1,922	1881	236,137-251,684	1922	1,401,948-1,440,361	1963	3,070,801-3,116,486
1841	1,923-2,412	1882	251,685-269,819	1923	1,440,362-1,478,995	1964	3,116,487-3,163,864
1842	2,413-2,900	1883	269,820-291,015	1924	1,478,996-1,521-589	1965	3,163,865-3,226,728
1843	2,901-3,394	1884	291,016-310,162	1925	1,521,590-1,568,039	1966	3,266,729-3,295,142
1844	3,395-3,872	1885	310,163-333,493	1926	1,568,040-1,612,699	1967	3,295,143-3,360,799
1845	3,873-4,347	1886	333,494-355,290	1927	1,612,700-1,654,520	1968	3,360,800-3,419,906
1846	4,348-4,913	1887	355,291-375,719	1928	1,654,521-1,696,896	1969	3,419,907-3,487,469
1847	4,914-5,408	1888	375,720-395,304	1929	1,696,897-1,742,180	1970	3,487,470-3,551,908
1848	5,409-5,992	1889	395,305-418,664	1930	1,742,181-1,787,423	1971	3,551,909-3,631,538
1849	5,993-6,980	1890	418,665-443,986	1931	1,787,424-1,839,189	1972	3,631,539-3,707,728
1850	6,981-7,864	1891	443,987-466,314	1932	1,839,190-1,892,662	1973	3,707,729-3,781,913
1851	7,865-8,621	1892	466,315-488,975	1933	1,892,663-1,941,448	1974	3,781,914-3,858,240
1852	8,622-9,511	1893	488,976-511,743	1934	1,941,449-1,985,877	1975	3,858,241-3,930,270
1853	9,512-10,357	1894	511,744-531,618	1935	1,985,878-2,026,515	1976	3,930,271-4,000,519
1854	10,358-12,116	1895	531,619-552,501	1936	2,026,516-2,066,308	1977	4,000,520-4,065,811
1855	12,117-14,008	1896	552,502-574,368	1937	2,066,309-2,104,003	1978	4,065,812-4,131,951
1856	14,009-16,323	1897	574,369-596,466	1938	2,104,004-2,142,079	1979	4,131,952-4,180,866
1857	16,324-19,009	1898	596,467-616,870	1939	2,142,080-2,185,169	1980	4,180,867-4,242,756
1858	19,010-22,476	1899	616,871-640,166	1940	2,185,170-2,227,417	1981	4,242,757-4,308,621
1859	22,477-28,841	1900	640,167-664,826	1941	2,227,418-2,268,539	1982	4,308,622-4,366,578
1860	28,842-31,004	1901	664,827-690,384	1942	2,268,540-2,307,006	1983	4,366,579-4,423,522
1861	31,005-34,044	1902	690,385-717,520	1943	2,307,007-2,338,080	1984	4,423,523-4,490,884
1862	34,045-37,265	1903	717,521-748,566	1944	2,338,081-2,366,153	1985	4,490,885-4,562,595
1863	37,266-41,046	1904	748,567-778,833	1945	2,366,154-2,391,855	1986	4,562,596-4,633,525
1864	41,047-45,684	1905	778,834-808,617	1946	2,391,856-2,413,674	1987	4,633,526-4,716,593
1865	45,685-51,783	1906	808,618-839,798	1947	2,413,675-2,433,823	1988	4,716,594-4,794,651
1866	51,784-6,0657	1907	839,799-875,678	1948	2,433,824-2,457,796	1989	4,794,652-4,890,334
1867	60,658-72,958	1908	875,679-908,435	1949	2,457,797-2,492,943	1990	4,890,335-4,980,926
1868	72,959-85,502	1909	908,436-945,009	1950	2,492,944-2,536,015	1991	4,980,927-5,077,835
1869	85,503-98,459	1910	945,010-980,177	1951	2,536,016-2,580,378	1992	5,077,336-5,175,885
1870	98,460-110,616	1911	980,178-1,013,094	1952	2,580,379-2,624,045	1993	5,175,886-5,274,845
1871	110,617-122,303	1912	1,013,095-1,049,325	1953	2,624,046-2,664,561	1994	5,274,846-5,377,358
1872	122,304-134,503	1913	1,049,326-1,083,266	1954	2,664,562-2,698,433	1995	5,377,359-5,479,657
1873	134,504-146,119	1914	1,083,267-1,123,211	1955	2,698,434-2,728,912	1996	5,479,658-5,590,419
1874	146,120-158,349	1915	1,123,212-1,166,418	1956	2,728,913-2,775,761	1997	5,590,420-5,704,061
1875	158,350-171,640	1916	1,166,419-1,210,388	1957	2,775,762-2,818,566	1998	5,704,062-
1876	171,641-185,812	1917	1,210,389-1,251,457	1958	2,818,567-2,866,972		

Table 2
United States Design Patents And Their Time Frames

Year	Number	Year	Number	Year	Number	Year	Number
1843	1-14	1882	12,647-13,507	1921	56,844-60,120	1960	186,973-189,515
1844	15-26	1883	13,508-14,527	1922	60,121-61,747	1961	189,516-192,003
1845	27-43	1884	14,528-15,677	1923	61,748-63,674	1962	192,004-194,303
1846	44-102	1885	15,678-16,450	1924	63,675-66,345	1963	194,304-197,268
1847	103-162	1886	16,451-17,045	1925	66,346-69,169	1964	197,269-199,954
1848	163-208	1887	17,046-17,994	1926	69,170-71,771	1965	199,955-203,378
1849	209-257	1888	17,995-18,829	1927	71,772-74,158	1966	203,379-206,566
1850	258-340	1889	18,830-19,552	1928	74,159-77,346	1967	206,567-209,731
1851	341-430	1890	19,553-20,438	1929	77,347-80,253	1968	209,732-213,083
1852	431-539	1891	20,439-21,274	1930	80,254-82,965	1969	213,084-216,418
1853	540-625	1892	21,275-22,091	1931	82,966-85,902	1970	216,419-219,636
1854	626-682	1893	22,092-22,993	1932	85,903-88,846	1971	219,637-222,792
1855	683-752	1894	22,994-23,921	1933	88,847-91,257	1972	222,793-225,694
1856	753-859	1895	23,922-25,036	1934	91,258-94,178	1973	225,695-229,728
1857	860-972	1896	25,037-26,481	1935	94,179-98,044	1974	229,729-234,032
1858	973-1,074	1897	26,482-28,112	1936	98,045-102,600	1975	234,033-238,314
1859	1,075-1,182	1898	28,113-29,915	1937	102,601-107,737	1976	238,315-242,880
1860	1,183-1,365	1899	29,916-32,054	1938	107,738-112,764	1977	242,881-246,810
1861	1,366-1,507	1900	32,055-33,812	1939	112,765-118,357	1978	246,811-250,675
1862	1,508-1,702	1901	33,813-35,546	1940	118,358-124,502	1979	250,676-253,795
1863	1,703-1,878	1902	35,547-36,186	1941	124,503-130,988	1980	253,796-257,745
1864	1,879-2,017	1903	36,187-36,722	1942	130,989-134,716	1981	257,746-262,494
1865	2,018-2,238	1904	36,723-37,279	1943	134,717-136,945	1982	262,495-267,439
1866	2,239-2,532	1905	37,280-37,765	1944	136,946-139,861	1983	267,440-272,008
1867	2,533-2,587	1906	37,766-38,390	1945	139,862-143,385	1984	272,009-276,948
1868	2,858-3,303	1907	38,391-38,979	1946	143,386-146,164	1985	276,949-282,019
1869	3,304-3,809	1908	38,980-39,736	1947	146,165-148,266	1986	282,020-287,539
1870	3,810-4,546	1909	39,737-40,423	1948	148,267-152,234	1987	287,540-293,499
1871	4,547-5,431	1910	40,424-41,062	1949	152,235-156,685	1988	293,500-299,179
1872	5,432-6,335	1911	41,063-42,072	1950	156,686-161,403	1989	299,180-305,274
1873	6,336-7,082	1912	42,073-43,414	1951	161,404-165,567	1990	305,275-313,300
1874	7,083-7,968	1913	43,415-45,097	1952	165,568-168,526	1991	313,301-322,877
1875	7,969-8,883	1914	45,098-46,812	1953	168,527-171,240	1992	322,878-330,169
1876	8,884-9,685	1915	46,813-48,357	1954	171,241-173,776	1993	332,170-342,817
1877	9,686-10,384	1916	48,358-50,116	1955	173,777-176,489	1994	342,818-353,931
1878	10,385-10,974	1917	50,117-51,628	1956	176,490-179,466	1995	353,932-365,670
1879	10,975-11,566	1918	51,629-52,835	1957	179,467-181,828	1996	365,671-377,106
1880	11,567-12,081	1919	51,836-54,358	1958	181,829-184,203	1997	377,107-388,584
1881	12,082-12,646	1920	54,359-56,843	1959	184,204-186,972	1998	388,585-

References

Articles

Bamford, Joan. "Strike a Light for Matchholders." *American Collector* 9, no.1 (Jan 1978): 36-38.

Books

Alsford, Denis B. *Match Holders*. Atglen, Pennsylvania: Schiffer Publishing Ltd., 1994.

Freeman, Larry. *Victorian Silver*. Watkins Glen, New York: Century House, 1967.

Godden, Geoffrey A., F .R. S. A. *Encyclopaedia Of British Pottery And Porcelain Marks*. New York: Bonanza Books, MCMLXV.

Kovel, Ralph and Terry. *New Dictionary of Marks*. New York: Crown Publishers, Inc., 1986.

McNerney, Kathryn. *Antique Iron*. Paducah, Kentucky: Collector Books, 1984.

Rainwater, Dorothy T., and H. Ivan. *American Silverplate*. Nashville, Tennessee: Thomas Nelson Inc.; Hanover, Pennsylvania: Everybodys Press, 1968.

Rainwater, Dorothy T. *Encyclopedia Of American Silver Manufacturers*. 3rd ed., rev. West Chester, Pennsylvania: Schiffer Publishing, Ltd.,1986.

Rimalover, Betty H. *Antique American Wall Match Holders*. Princeton Junction, New Jersey: Stonybrook Associate, 1976.

Schnaldig, Victor K. *American Victorian Figural Napkin Rings*. Demoines, Iowa: Wallace Homestead Book Co., 1971.

Sullivan, Audrey G. *A History Of Match Safes In The United States*. Ft. Lauderdale, Florida: Riverside Press, Inc., 1978.

Brochures

Starting From Scratch. New York: Diamond National Corporation, Diamond Match Division, 1962.

The Past Is Prologue: First Hundred Years 1881-1981. Minneapolis, Minnesota.

Diamond International Corporation.

Patents And Their Time Frames. San Jose, California: United States Patent Office.

Manchester, Herbert. *Fifty Years of Match Making: The Diamond Match Company*. New York: Blanchard Press, Inc.,1928.

Catalogues

Marshal Field & Co. 1896 Catalogue Reproduction. Northfield, Illinois: Gun Digest Publishing Company,1970.

The Peter Sidlow Collection. New Hope, Pennsylvania: Noel Barrett Antiques and Auctions Presents, 1989.

Pairpoint Manufacturing Co., 1894 Catalogue Reproduction. Washington Mills, New York: The Gilded Age Press, 1979.

INDEX